D1327012

Predictioneer

*One who uses maths, science and
the logic of brazen self-interest
to see and shape the future*

Predictioneer

One who uses maths, science and the logic of brazen self-interest to see and shape the future

BRUCE BUENO DE MESQUITA

THE BODLEY HEAD
LONDON

Coventry City Council	
CEN	
3 8002 01726 876 6	
Askews	Sep-2009
303.49	£16.99

Published by The Bodley Head 2009

2 4 6 8 10 9 7 5 3 1

Copyright © Bruce Bueno de Mesquita 2009

Bruce Bueno de Mesquita has asserted his right under the Copyright, Designs and Patents Act 1988 to be identified as the author of this work.

This book is sold subject to the condition that it shall not, by way of trade or otherwise, be lent, resold, hired out, or otherwise circulated without the publisher's prior consent in any form of binding or cover other than that in which it is published and without a similar condition, including this condition, being imposed on the subsequent purchaser.

First published in Great Britain in 2009 by
The Bodley Head
Random House, 20 Vauxhall Bridge Road,
London SW1V 2SA

www.bodleyhead.co.uk
www.rbooks.co.uk

Addresses for companies within The Random House Group Limited can be found at: www.randomhouse.co.uk/offices.htm

The Random House Group Limited Reg. No. 954009

A CIP catalogue record for this book is available from the British Library

ISBN 9781847920669
ISBN 9781847920676

The Random House Group Limited supports The Forest Stewardship Council (FSC), the leading international forest certification organisation. All our titles that are printed on Greenpeace approved FSC certified paper carry the FSC logo. Our paper procurement policy can be found at: www.rbooks.co.uk/environment

Mixed Sources
Product group from well-managed forests and other controlled sources
www.fsc.org Cert no. TT-COC-2139
© 1996 Forest Stewardship Council

Printed and bound in Great Britain by
Clays Ltd, St Ives PLC

For my grandchildren,
Nathan, Clara, Abraham, Hannah,
and those who may be yet to come.
They will be fine caretakers
of the future.

Contents

Introduction

◼

KING LEOPOLD II, remembered today as Belgium's Builder King, reigned from 1865 to 1909.[1] A constitutional monarch who, like many of his contemporaries, longed for the bygone days of absolute power, he was nonetheless an unusually influential and activist king who helped make Belgians free, prosperous, and secure.

Belgium's good works during Leopold's reign are almost uncountable. He oversaw the expansion of political freedom with the adoption of universal adult male suffrage in competitive elections, putting his country on a firm footing to become a modern democracy. On the economic front, he encouraged free-trade policies that guided Belgium to remarkable growth. In little Belgium, coal production, the engine of industry in nineteenth-century Europe, rose to such heights that it almost equaled that of France. Social policy too moved briskly ahead. Primary education became compulsory, and with the 1881 School Law, girls were assured access to secondary education.[2] Moreover, Leopold's policies provided greater protection for women and children than was then the norm in most of Europe. Thanks to legislation passed in 1889, children under twelve could not be put to work, and after they turned twelve their workdays were limited to twelve hours, a radical departure from prevailing policy of the time.

When the Belgian economy was racked by a major economic crisis in 1873, Leopold helped improve the lot of the poor with pro-labor reforms, including granting workers the right to strike, a right that was still hotly resisted in the United States half a century later. He promoted truly ambitious public works projects, including massive road and railway construction designed to reduce unemployment, promote urbanization, and increase business opportunities. He was way ahead of Franklin Delano Roosevelt and Barack Obama in recognizing how to stimulate employment and economic prosperity by building up infrastructure.

Leopold was a great reformer at home, a founder of Belgium's long years of peace and plenty.

But then there was the Congo.

Though he never set foot in Africa, Leopold also ruled over the Congo Free State for nearly a quarter of a century (1885–1908). He built his personal wealth in the Congo first by extracting high-priced ivory from the region and then by exploiting the even more lucrative rubber trade that developed there. Unlike in Belgium, there was no *chef de cabinet* (roughly, prime minister), and no voters among the Congo's approximately 30 million people to limit what he could do. Because it was his personal property, Leopold was free to exert the absolute rule he could not have at home. His "police," the Force Publique, became the key to governing the Congo. Their job was to extract wealth for him (and for themselves) by ensuring the vast exportation of rubber to meet world demand. This band of butchers was led by a small number of Europeans who kidnapped and enslaved Congolese as soldiers who were in turn responsible for making sure that rubber quotas were met. Slave labor was the Force Publique's preferred mode of production, so its soldiers set about enslaving Congolese men, women, and children.

Leopold's "police" received low salaries but could earn big commissions by meeting or exceeding their rubber quotas. Unrestricted by any law governing their conduct except, literally, the law of the jungle, and provided with a huge financial incentive through the commission system, these soldiers of sorrow, from the very bottom of the ladder to the very top, used whatever means they saw fit to meet the quotas. The incentives included not only riches for those who succeeded but the severest punishment for those who failed, including beatings and even death. To avoid this fate the police tortured, maimed, and often murdered those below them who threatened (or could be claimed to have threatened) rubber production. Rewarded for killing people allegedly engaged in antigovernment activities and needing to account for every bullet they spent, soldiers quickly took to indiscriminate mutilation of innocent souls as a way to boost their counts and thereby their fees, going so far as to chop off the right hands of women and children to provide evidence of their work on behalf of Leopold's interests. Perhaps as many as 10 million people were murdered at the hands of the Force Publique in their pursuit of wealth for Leopold and, of course, for themselves.[3]

In contrast to Leopold's progressive policies in Belgium, virtually nothing was invested in improving conditions in the Congo. Roads were built only where they helped move rubber to market. Laws protecting women and children or worker's right to strike were unheard of. Much as Leopold worried about protecting the security of his Belgian subjects, he worked to undermine the security of his Congolese subjects. Just about the only items exported to the Congo were weapons for the Force Publique, while vast riches flowed back to Europe. Indeed, it was this extraordinary imbalance in trade that eventually led to the revelation in Belgium that Leopold was growing rich through slavery and much worse. In 1908, the evidence of atrocities reached such a level that they could no longer be denied, and Leopold, with great reluctance, surrendered his control over the Congo to the Belgian government. The ministers certainly did not rule it well, but compared to Leopold, they were a significant improvement.

How could King Leopold II have ruled two places at the same time in such dramatically different manners?

It's easy to blame Leopold's apparent split personality—a progressive in Belgium and a monster in the Congo—on some character flaw or on a diseased mind. It's also easy to explain away his horrible rule in the Congo as typical racist behavior. These explanations feel good, but they almost certainly cannot describe the big picture. After all, just think about Mobutu Sese Seko, the Congo's latter-day Leopold, the monster in a leopard-skin hat who ruled Zaire (largely what used to be the Congo Free State and is today the Democratic Republic of Congo) for more than thirty years (1965–97). During that time he bankrupted his country, stole billions of dollars for himself, and murdered hundreds of thousands of Congolese. Surely we cannot blame Mobutu's murderous rule on racism. Was he crazy? Probably not, and besides, what are the odds that so many allegedly crazy people would rise to and then successfully cling to power for decades despite their awful rule?

Leopold and Mobutu are far from unusual cases. Even today, the United Nations reports that people caught up in Sierra Leone's diamond war have had their hands and feet cut off. Similar policies of mutilation, torture, and murder are reported in Zimbabwe and occurred in the genocide in Rwanda. And then we should not forget the Holocaust or, more recently, Cambodia's Pol Pot, who ordered the murder of millions of Cambodians for such crimes as wearing eyeglasses (proof that they were educated and therefore probably a threat to the regime). Such monstrous rulers are not a thing of the past. Murder and misery have been mainstays of long-lasting leaders throughout history, a fact that remains as true today as a hundred or a thousand years ago.[4]

It's nice to think that leaders who provide peace and plenty rule for long, happy years, beloved by the people and content to do good for them day and night. But in fact those who want to run a country for a long time are ill advised to go around promoting peace and prosperity. Not that making people well off is inherently bad for leaders; it isn't. It's just that promoting corruption and misery is *better*. That was well understood by Leopold and Mobutu in the Congo, and is clearly understood today by the governments in places like North Korea, Zimbabwe, Turkmenistan, Chad, Syria . . . sadly, the list goes on.

It so happens that leaders who are really good at giving their people life, liberty, and happiness are, overwhelmingly, democratically elected and therefore face organized political competition. It also so happens that they are routinely thrown out after only a short time in office.

It's true that Leopold ruled Belgium for forty-four years, but he was a constitutional monarch who had to work within the constraints of the democratic system that governed Belgium if he was to remain in power. Yet in looking at modern democratic governments, we see that doing right by the people is no guarantee of political longevity. During Golda Meir's period as prime minister, Israel enjoyed a 9 percent average annual growth rate. She held office for just four years. Japan's Eisaku Sato presided over a 9.8 percent growth rate, surviving as prime minister for less than eight years. Perhaps most famously, in 1945, after five years in office and less than two months after Germany's surrender in World War II, Winston Churchill was tossed out as prime minister of Great Britain and replaced by Clement Attlee, despite having (allowing for slight exaggeration) saved the United Kingdom itself.

Why, in contrast, do those leaders who make their subjects' lives miserable typically die in their sleep or live out their retirement years lounging on a luxurious beach after being in office twenty, thirty, or forty or more years? It's my claim, and it may seem controversial, that kleptocratic leaders are not inherently evil—at least not necessarily so—and that those who do a great job for their people in hopes of reelection are hardly fit for sainthood. They're all doing the right things if they want to stay in power as long as possible. Leopold, despicable as he was, did what worked best for him in the politically unconstrained environment of the Congo, and he did what worked best for him in the constitutionally limiting environment of Belgium.

The difference between doing a good job and doing a lousy job is driven by how many people a leader *has* to keep happy. Why doesn't every leader allow cronies to loot and steal the way the Force Publique did? Large-scale democratic leaders can't—they have to reward too many people to make theft and corruption work for them. In other words, the system does not effectively incentivize that strategy. Virtually all long-lasting (read authoritarian) leaders, however, really depend only on a very small number of generals, senior civil servants, and their own families for support. Because they rely on so few people to keep them in power, they can afford to bribe them handsomely. With such big paydays, those cronies aren't going to risk losing their privileges. They'll do whatever it takes to keep the boss in power. They will oppress their fellow citizens; they'll silence a free press and punish protesters. They will torture, maim, and murder to protect the incumbent as long as the incumbent delivers enough goodies to them.

The rub is that even when crony-dependent leaders want to do good deeds, they dare not pay for them with money promised to their essential supporters. Taking money from their cronies' pockets is a sure way to get overthrown. Spend too much on helping the people, and the cronies will find someone new to take over the top spot, someone who will pay them reliably instead of "dissipating" money on the masses.[5]

Like autocrats, elected officials are held accountable by people who want to know, *What have you done for me lately?*—except, for elected leaders, there are millions of such potential backers (or detractors, if not made happy), as opposed to hundreds. Democratic leaders *have* to act as if they care about the masses. Their campaigns are always an arms race in policy ideas: which candidate has (or appears to have) the best ideas about health care, about taxes, about national security, about education, and on and on. When a seemingly fit democratic leader is thrown out of office, it's generally because his or her opponent is perceived to be just a little bit better—a remarkably positive condition, particularly when the alternatives are considered.

So the explanation for Leopold the Builder King and Leopold the Monster has begun to fall into place. When rulers need the support of many—as was Leopold's situation in Belgium—the best way to rule is by creating good policies. When leaders rely only on a few to stay in control—as was the case for Leopold in the Congo—their best bet is to make the few fat and happy, even if that means making everyone else miserable. But let's take this a step further.

Leopold, Mobutu Sese Seko, and Golda Meir were all powerful leaders, but (to state a most obvious but important fact) they were all people, too, and no different from the rest of us. Whether in government or business, we all want to keep our jobs, we all seek advantage in the accumulation of wealth or influence, and we all evaluate our self-interest, often ahead of such lofty ideas as the national interest or notions of corporate well-being.

With this in mind, if we were to turn back the clock, could we not have made some educated predictions that Leopold, the very same man, would behave differently as the head of the Belgian and Congolese states? Could we not have surmised that Mobutu Sese Seko would rule in the fashion that he did? Or that Churchill would lose power when the attention of the British people turned to postwar reconstruction and domestic matters? Or, in a completely different setting, could we see how a corporate partnership's structure might encourage its members to overlook fraud? And wouldn't knowing these things ahead of time be of some potential value?

I believe the answer to all of those questions is yes, which brings me to the very purpose of my work and to the principal claim of this book: that it is possible for us to anticipate actions, to predict the future, and, by looking for ways to change incentives, to engineer the future across a stunning range of considerations that involve human decision making. That's not to say it's easy, or that it's a mere matter of anecdote and reflection—there's hard science, theory, and some mind-bending arithmetic that come into play—but it is possible, and given what we've seen when humans in power run amok, whether in châteaus or boardrooms, it's preferable to letting the chips fall where they may or to saving our better ideas, regrets, and outrage for when they usually appear—that is, when they're too late.

■   ■   ■

Who am I that you should care what I think about these big questions? And why in the world should you take me seriously as a predictioneer?

It so happens I've been predicting future events for three decades, often in print before the fact, and mostly getting them right. Don't get me wrong—I'm no soothsayer and I have no patience for crystal ball gazers, astrologers, or even most pundits. In my world, science, not mumbo-jumbo, is the way to anticipate people's choices and their consequences for altering the future. I use game theory—we'll talk later about what that means—to do just that for the U.S. government, big corporations, and sometimes ordinary folks too. In fact, I have made hundreds, even thousands of predictions—a great many of them in print, ready to be scrutinized by any naysayer. There is nothing uncanny about my ability to predict. Anyone can learn to use scientific reasoning to do what I do, and I'm going to show you a bit of how to do it right here. But first, let me fill you in a little bit about how I got into the prediction business.

I'm a political science professor at New York University, where I also run the Alexander Hamilton Center for Political Economy. The Center and all of my courses try to teach students how to solve problems with logic and evidence. The idea is to wean them from knee-jerk conclusions based on gut feel, personal opinions, simple linear reasoning, partisan preferences, or ideology. My colleagues at NYU and I are interested in training students to know how to address problems before they go out into the world. We don't want them shaking things up without much insight into whether they're helping to make matters better or worse.

Besides being a professor at NYU, I wear two other hats. I'm a senior fellow at the Hoover Institution at Stanford University. There my job is to think about finding solutions to policy problems. That side of my research is about putting the ideas I teach at NYU to good use by writing op-ed columns, articles, and books, some very technical and some, like this one, designed to spread the word. My third hat is as a partner in a small consulting company, Mesquita & Roundell, LLC. M&R, as we call it, also uses some of the game theory models I've designed to advise people in the national security community and also in the private sector.

I didn't set out to wear these three hats. The opportunity initially fell into my lap back in 1979 when an official at the State Department called me to ask my opinion about a government crisis in India. He wanted to know who was likely to be the next prime minister. At the time I was a professor of political science at the University of Rochester—where the application of game theory to political questions originated—and I had written my Ph.D. thesis at the University of Michigan about winning and losing strategies among India's opposition parties. So the State Department official was asking me to be a pundit, to use my "expert" knowledge to speculate about the next Indian government.

It happened that I was on a Guggenheim fellowship at the time, working on a book about war. I had just designed a mathematical model for that project, as well as a little computer program to make calculations that were important for solving that model. The computer program provided a way to simulate decision making under stressful circumstances such as sometimes lead to war. It looked at the choices people could make and calculated the probability that they would get what they wanted if they chose one course of action (say, negotiations) or another (like war), weighting the probabilities by an estimate of how much the decision makers valued winning, losing, or intermediate compromise outcomes. Of course, it also recognized that they had to work out how others might respond to the choices they made.

Like every model, it needed data. The State Department's phone call about India came in just as I was trying to figure out where to get data to feed into my war and peace model. The timing was perfect. The phone call got me thinking that maybe war and peace decisions really aren't that different from everyday political confrontations. Sure, the stakes are higher—people get killed in wars—but then any politician seeking high office or about to lose high office sees the personal political stakes as pretty darn high. Probably all of us make similar calculations about how to advance our own well-being in any complex situation involving big risks and potentially big rewards, whether that involves politics, business, or daily life.

The State Department was pressing me for an answer and I wanted to help them. I also wanted to see how well my new model worked. I decided to find out whether the model could really be a useful tool to sort out the political infighting in India. Linking that model to Indian politics was a huge "Aha!" moment for me, one that would change the rest of my life.

I grabbed a yellow pad and picked my own brain, putting together the information the model needed. I wrote down a list of everyone I thought would try to influence the selection of India's next government. For each of those people (political party leaders, members of India's parliament, and some members of critical state governments) I also wrote down my estimate of how much clout each had, what their preference was between the various plausible candidates for prime minister, and how much they cared about trying to shape that choice. With just one page of my yellow pad filled with numbers, I had all the information the computer program needed to predict what would happen. I plugged those data into my little program and crunched the numbers overnight. When the computing was done the next morning—computers were slow in those days—I pored over the hundred or so pages of calculated values to see what the model's predictions looked like.

I thought I had personal insight into what was going to happen in India. My "pundit" knowledge had led me to believe that a man named Jagjivan Ram would be the next prime minister. He was a popular and prominent politician who was better liked than his main rivals for the prime minister's job. I was confident that he was untouchable—truly unbeatable—in the political arena, and not just in the sense of his caste status. He had paid his political dues and it seemed like his time had come. Many other India watchers thought the same thing. Imagine my surprise, then, when my computer program, written by me and fed only with my data, predicted an entirely different result. It predicted that Charan Singh would become prime minister and that he would include someone named Y. B. Chavan in his cabinet, and that they would gain support—albeit briefly— from Indira Gandhi, then the recently ousted prime minister. The model also predicted that the new Indian government would be incapable of governing and so would soon fall.

I found myself forced to choose between personal opinion and my commitment to logic and evidence as the basis for coming to conclusions about politics. I believed in the logic behind my model and I believed in the correctness of the data I had jotted down. After staring at the output, working out how my own program came to a conclusion so different from my personal judgment, I chose science over punditry. In fact, I told colleagues at Rochester what the model's prediction was even before I reported back to the State Department. When I spoke with the official at State he was taken aback. He noted that no one else was suggesting this result and that it seemed strange at best. He asked me how I had come to this judgment. When I told him I'd used a computer program based on a model of decision making that I was designing, he just laughed and urged me not to repeat that to anyone.

A few weeks later, Charan Singh became the prime minister with Y. B. Chavan as his deputy prime minister, with support from Indira Gandhi. And a few months after that, Charan Singh's government unraveled, Indira Gandhi withdrew her support, and a new election was called, just as the computer-generated forecast had indicated. This got me pretty excited. Here was a case where my personal judgment had been wrong, and yet my knowledge was the only source of information the computer model had. The model came up with the right answer and I didn't. Clearly there were at least two possibilities: I was just lucky, or I was onto something.

Luck is great, but I'm not a great believer in luck alone as an explanation for results. Sure, rare events happen—rarely. I set out to push my model by testing it against lots of cases, hoping to learn whether it really worked. I applied it to prospective leadership changes in the Soviet Union; to questions of economic reform in Mexico and Brazil; and to budgetary decisions in Italy—that is, to wide-ranging questions about politics and economics. The model worked really well on these cases—so well, in fact, that it attracted the attention of people in the government who heard me present some of the analyses at academic conferences. Eventually this led to a grant from the Defense Advanced Research Projects Agency (DARPA), a research arm of the Department of Defense (and the sponsors of research that fostered the development of the Internet long before Al Gore "invented" it). They gave me seventeen issues to examine, and as it happened, the model—by then somewhat more sophisticated—got all seventeen right. Government analysts who provided the data the model needed—we'll talk more about that later—didn't do nearly as well. Confident that I was onto something useful, I started a small consulting company with a couple of colleagues who had their own ideas about how to predict big political events. Now, many years later, I operate a small consulting firm with my partner and former client, Harry Roundell. Harry, formerly a managing director at J. P. Morgan, and I apply a much more sophisticated version of my 1979 model to interesting business and government problems. We'll see lots of examples in the pages to come.

It's easy to see if predictions are right or wrong when they are precise, and almost impossible to judge them when they are cloaked in hazy language. In my experience, government and private businesses want firm answers. They get plenty of wishy-washy predictions from their staff. They're looking for more than "On the one hand this, but on the other hand that"—and I give it to them. Sometimes that leads to embarrassment, but that's the point. If people are to pay attention to predictions, they need real evidence as to the odds that the predictions are right. Being reluctant to put predictions out in public is the first sign that the prognosticator doesn't have confidence in what he's doing.

According to a declassified CIA assessment, the predictions for which I've been responsible have a 90 percent accuracy rate.[6] This is not a reflection of any great wisdom or insight on my part—I have little enough of both, and believe me, there are plenty of ivy-garlanded professors and NewsHour intellectuals who would agree. What I do have is the lesson I learned in my "Aha!" moment: Politics is predictable. All that is needed is a tool—like my model—that takes basic information, evaluates it by assuming everyone does what they think is best for them, and produces reliable assessments of what they will do and why they will do it. Successful prediction does not rely on any special personal qualities. You don't need to walk around conjuring the future, plucking predictions out of thin air. There's no need for sheep entrails, tea leaves, or special powers. The key to good prediction is getting the logic right, or "righter" than any way that is achieved by other means of prediction.

Accurate prediction relies on science, not artistry—and certainly not sleight of hand. It is a reflection of the power of logic and evidence, and testimony to the progress being made in demystifying the world of human thought and decision. There are lots of powerful tools for making predictions. Applied game theory, my chosen method, is right for some problems but not all. Statistical forecasting is a terrific way to address questions that don't involve big breaks from past patterns. Election prognosticators, whether at universities, polling services, or blogs on the Web (like Nate

Silver, the son of an old family friend) all estimate the influence of variables on past outcomes and project the weight of that influence onto current circumstances. Online election markets work well too. They work just the way jelly bean contests work. Ask lots of people how many jelly beans there are in a jar, and just about no one will be close to being right, but the average of their predictions is often very close to the true number. These methods have terrific records of accuracy when applied to appropriate problems.

Statistical methods are certainly not limited to just studying and predicting elections. They help us understand harder questions too, such as what leads to international crises or what influences international commerce and investments. Behavioral economics is another prominent tool grounded in the scientific method to derive insights from sophisticated statistical and experimental tests. Steven Levitt, one of the authors of *Freakonomics,* has introduced millions of readers to behavioral economics, giving them insights into important and captivatingly interesting phenomena.

Game-theory models, with their focus on strategic behavior, are best for predicting the business and national-security issues I get asked about. I say this having done loads of statistical studies on questions of war and peace, nation building, and much more, as well as historical and contemporary case studies. Not every method is right for every problem, but for predicting the future the way I do, game theory is the way to go, and I'll try to convince you of that not only by highlighting the track record of my method, but also by daring to be embarrassed later in this book when I make predictions about big future events.

Prediction with game theory requires learning how to think strategically about other people's problems the way you think about your own, and it means empathizing with how others think about the same problems. A fast laptop and the right software help, but any problem whose outcome depends on many people and involves real or imagined negotiations is susceptible to accurate forecasting drawn from basic methods.

In fact, not only can we learn to look ahead at what is likely to happen, but—and this is far more useful than mere prediction and the visions of seers past and present—we can learn to engineer the future to produce happier outcomes. Sadly, our government, business, and civic leaders rarely take advantage of this possibility. Instead, they rely on wishful thinking and yearn for "wisdom" instead of seeking help from cutting-edge science. In place of analytic tools they count on the ever-present seat of their pants.

We live in a world in which billions—even trillions—of dollars are spent on preparations for war. Yet we spend hardly a penny on improving decision making to determine when or whether our weapons should be used, let alone how we can negotiate successfully. The result: we get bogged down in far-off places with little understanding of why we are there or how to advance our goals, and even less foresight into the road-blocks that will lie in our way. That is no way to run a twenty-first-century government when science can help us do so much better.

Business leaders do no better than their political peers. They spend for-tunes doing financial analyses of their expected gains and losses from this or that deal, but they spend virtually nothing analyzing how their coun-terparts on the other side of the table think about their own gains and losses. The result: companies have a good idea how much a business is worth to them before they try to buy it, but they don't know how much they need to pay for it. In my experience, they often pay far too much, or, to look at the other side of the coin, they sell for much less than the buyer was prepared to pay. Too bad for their shareholders.

How can anyone make prudent choices without first thinking through how others will see those choices and react to them? Yet that is how most big decisions are made, blind to anyone else's point of view. Plowing ahead without much thought to what motivates our rivals, whether in business or in government, is a surefire way to make a mess of things, leaving us to muddle through at best, our hopes pinned to shortsighted decisions.

Decision making is the last frontier in which science has been locked out of government and business. We live in a high-tech age with archaic guesswork guiding life and death decisions. The time for peering into tea leaves or reading astrological charts should be long over. We should leave entertaining divinations to storefront psychics and open the door to sci-ence as the new basis for the big decisions of our time.

Curious about how this can be done? The chapters to follow explain how precise predictions can be made. We will see, through illustrative examples from the worlds of national security and business and every-day life, that the problems of war and peace, mergers and acquisitions, litigation, legislation, and regulation—and just about anything else that does not rely on the hidden hand of market forces—can be reliably predicted.

We will see ways to use science, mathematics, and, in particular, the power of game theory to sort out behavior and improve the future. I hope to share with you this cutting-edge world of thought, whose potential, to many, may seem to verge on the mystical. But there is no mystery or mysticism in good prediction. To demonstrate this for you, I will suggest, in Chapter One, how a modest amount of strategic reasoning can help you save hundreds and maybe thousands of dollars the next time you buy a car.

Predictioneer

1

■

WHAT WILL IT TAKE TO
PUT YOU IN THIS CAR TODAY?

"GAME THEORY" IS a fancy label for a pretty simple idea: that people do what they believe is in their best interest. That means they pay attention to how others might react if they choose to do one thing or another. Those "others" are anyone believed to be a prospective supporter or opponent. Looking at how their interests intersect or collide is basic to assessing potential outcomes of decision making. To get a good grip on what people are likely to do requires first approximating what they believe about the situation and what they want to get out of it. By estimating carefully people's wants and beliefs, anyone can make a reliable forecast of what each and every one of them will do. And if you can predict what will happen, then you can also predict what will happen if you alter what people believe about a situation. This is, in short, how we can use the same logic both for prediction and for engineering the future.

I'll provide a more detailed examination of game theory in the next two chapters, but first, as promised, let me illustrate what I'm talking about with the example of how best to buy a new car.

A new car purchase is a costly bargaining experience for most of us because most of us are bad at it. A little strategic reasoning can go a long way to improve the experience. If you follow the ideas here, you will not only have a happier time buying a car, you'll also pay a lot less money.

New cars are mostly bought in one of two ways. Most of us go to a dealership, test-drive a vehicle, maybe fall in love with it, and engage in a most unpleasant negotiation over the price. A smaller number of us hate that experience enough that we buy cars through the Internet. This usually means getting price bids from a few local dealers and then going to one of them to get the car. That's slightly better, but there's yet a better way to buy cars that I urge you to try.

What's wrong with the most popular means of car shopping? Pretty much everything. To start with, you, the buyer, invest your time, and probably the time of some family members as well, at a car dealership. The salesperson knows that few people enjoy dealing with them. They know that they rate near the bottom of anybody's scale of occupational trustworthiness. (The *Jobs Rated Almanac* weights occupations by numerous factors and reports that auto sales ranks 220th out of 250 occupations.[1] Apparently it's worse to be a taxi driver, cowboy, or roustabout, but not by much.) But there you are subjecting yourself to the salesperson's pitch, standing before a car dealer in his or her place of business, feeling compelled to haggle over the price, probably to your embarrassment and certainly to your disadvantage. The whole time you're talking to the dealer you're revealing information that sets you up to pay too much.

Being there is what game theorists call a "costly signal." It's a costly signal because your expenditure of time and energy announces that you want to buy, that there's a good chance you'll buy from the dealership you're visiting rather than go elsewhere, and, especially if you have kids with you, that you want to get out of there as quickly as possible. That first step, then, *your simply being there,* translates into strengthening the salesperson's hand in getting a good price. They believe you're ready to buy, and you've done precious little to dissuade them. Score one for the dealer, none for you.

Now as it happens, costly signals are usually good things for you. They show you're serious about what you are saying and doing. That can give you credibility, as we will see in later chapters. Unfortunately, costly signals can have just the opposite effect when you're a shopper. They announce your eagerness to buy, and that makes it tough to get a good deal.

The situation was bad enough when you walked in the dealership door, but it only gets worse once a conversation begins. Although you've probably done homework online and know something about the invoice price of the car you want, there's a great deal you don't know that the dealer does know. When you say you want a gray car, or a blue one, or yellow, you don't know whether you've picked the hottest color in the greatest demand or a color that hardly anyone wants. You may not even know that there can be a price difference of many hundreds of dollars between choosing, say, red or yellow because the dealer treats that choice just the way options are treated—as one more place to pile on costs. You don't know enough about the local supply and demand to have a good retort when Pat, the salesperson, tells you the vehicle you want is in short supply. Translation: The invoice price on the Internet? Forget it—that's not going to happen for the car you want. Score another point for the dealer, and still none for you.

As anyone who's ever bought a new car knows, Pat will almost certainly ask you, "What will it take to put you in this car today?" Now you're cornered. If you name a really low price, Pat, who has been playing up to you, looks hurt or loses interest, or maybe even becomes just a little rude. Control of the discussion is now fully out of your hands. You feel bad about the lowball offer, you want to establish your good faith, and so you come back with a number close to what you're actually prepared to pay because you want to keep the conversation going. So Pat has gotten you to put a credible price on the table. Now, after praising the offer, Pat announces that this is a price that can be taken to the sales manager for approval. This is when the price gets jacked up a little bit more to close the deal. Rather than worrying about what the sales manager will say as you sit at Pat's desk, sipping dishwater-thin coffee from a paper cup, ask yourself, *Is it possible that Pat does not know the vehicle's price?!* Hard to believe, since that's what Pat does all day—sell cars.

You have now given Pat almost all the important information that was in your hands. Meanwhile, you have learned very little about the true state of the market. You may believe that Pat is working on your behalf with the sales manager, telling the manager what a good customer you are and how nice your family is. My guess is that they talk about last night's football game or how lousy that coffee is, and then Pat emerges with a price.

After some haggling over minor details, hands are shaken and a deal is struck. You feel efficacious. You got a "good" price that you will boast about to your friends. Pat assures you that this is so—and maybe it is, but then the nature of any sale is that the buyer and the seller are both happy with the price; otherwise there's no deal. You really do not know that you've gotten the best price for the car, just as the salesperson does not know that you paid the most you were prepared to pay. But Pat controlled the agenda, revealed little, and worked your price up. That is one more point for the dealer and maybe, just maybe, one for you too. The winner: the dealer by at least 3 or 4 to 1.

Now, the Internet *can* work much better. There is lots of useful information to be had, including information that car dealers don't want buyers to know. It's a great source for finding out a car's invoice price, for example, and Internet car purchase services generally guarantee that you'll be contacted by two or three dealers who at least know that you've already done some online comparison shopping. Still, making a deal requires going to a dealership. Once you are in the dealership, everything that happens in the first way of buying cars is now going to happen again, except that Pat already knows you're interested at the price quoted online. Maybe Pat says the exact car you asked about isn't available. Sure, it could be special-ordered, but that will take a long time. Maybe Pat offers you a seemingly good deal on a different car and the information gained on the Internet fades in importance. How about a similar car with a few extras—maybe a sunroof, a super-duper sound system, special trim, whatever—that you really don't want? At least that car is on the dealer's lot, if only for just a little bit more money. Or maybe you close the deal on the car you asked about at the price quoted online. Well, you're probably *closer* to the lowest price the dealer will take, but in all likelihood you are not there. After all, they don't quote a price to you that they expect you to bid up from; they anticipate you're going to look for ways to cut their price. It must have some give left in it.

So, how else can you buy a car? Here's what I recommend so that information flows to you and very little information flows the other way; so that beliefs change in your favor and not in favor of the seller; so that prospective sellers are compelled by self-interest to reveal their lowest price (their reservation price, as economists call it) and you are not.

It is always advantageous to control the agenda in a negotiation. In the car-buying context, that means constructing the conversation so that the seller puts her best price on the table knowing that it will not be accepted right now no matter how good it is. To achieve this, the seller needs to be told the game you are playing, including that you're serious about buying a car. The game goes like this:

Do your homework and decide exactly what car you want to buy. You may go to dealerships for this purpose—I do not, it's too easy to get sucked in—or you may search online to find the car that best fits your needs. Decide what you value. Is it safety over performance? Style over comfort? Is the color important enough that you'll pay a premium for one rather than another? Know what options you want and how the options are bundled (sports package, safety package, sound system, etc.). When you know what car you want, including the options, the color, the model, everything, then and only then let your fingers do the walking. Find every dealer that sells the car brand you want within a radius of maybe twenty to fifty miles of your home. For most of us, urban dwellers that we are, that will be a goodly number of dealerships.

Telephone each dealership in the radius you picked. Don't worry about having to bring the car to a distant dealer for servicing. The manufacturer provides the warranty, not the dealer. Ask to be connected to a salesperson and tell the person precisely what you are doing. Here is my typical spiel: "Hello, my name is Bruce Bueno de Mesquita. I plan to buy the following car [list the exact model and features] today at five P.M. I am calling all of the dealerships within a fifty-mile radius of my home and I am telling each of them what I am telling you. I will come in and buy the car today at five P.M. from the dealer who gives me the lowest price. I need to have the all-in price, including taxes, dealer prep [I ask them not to prep the car and not charge me for it, since dealer prep is little more than giving you a washed car with plastic covers and paper floormats removed, usually for hundreds of dollars], everything, because I will make out the check to your dealership before I come and will not have another check with me."

If you are making your first call, be sure to tell the salesperson that you will tell the next dealer the price you've been quoted. After the first call, make sure the future salespeople know that you will be repeating whatever is the lowest price offered to you so far. That way, future dealers know what price they must beat, and the dealer you are currently talking to knows that if he wants to have a shot at selling you a car, he had better quote his lowest price. Pat's hands are really tied. Trying to eke out a higher price just means a worse chance of selling you a car. That advantage hardly ever arises when you go to a dealer and give Pat complete control. Instead, you have set up an auction in which everyone knows that they have one chance to make the best bid.

Dealers don't like getting these phone calls. Their typical first response is, "Sir, you cannot buy a car over the phone." My reply: "Well, I have purchased many cars this way, so maybe I can't buy a car from *you* over the phone, but I know I can from others. So if you do not want to sell a car to me, that is fine." That is the end of the conversation with a few dealers, but not many. The common follow-up is a line salespeople like to use on the showroom floor when you tell them you are shopping around. "Sir, if I quote a price to you, the next dealer will quote fifty dollars less and you'll buy it from him." Pat has reverted to the standard approach and is now trying to regain control by making me feel I should be willing to pay an extra $50, as if I owe at least that to good old Pat. My response: "That's right. I'll buy it from the other dealer. So, Pat, if you can quote me that fifty dollars less, this is your opportunity to do so." Then they often say, "Believe me, come in, we have the best prices in town." The reply, "Good, then you should be happy to quote the price to me now because you are confident it is the best price. The only reason not to tell me the price on the phone is because you think it will be beaten." If the conversation has continued—and it does with a substantial portion of dealerships—Pat will quote a price.

I arrive at the lowest-priced dealer, check in hand, just before 5:00 P.M. to close the deal. If there is any change in the terms I leave immediately and go to the second-best offer, and so forth. I have only once had to pay the second-best price quoted to me.

I have found that the quoted prices vary tremendously from dealer to dealer, literally by thousands of dollars. I have personally purchased Toyotas, Hondas, and a Volkswagen this way. Some of my students at NYU have taken up this method and bought cars this way too. It's a nice little payback for their huge tuition bill. They and I have always beaten the price quoted on the Internet with this method. I even purchased a car for my daughter this way when I was three thousand miles from the dealership (she was going in at 5:00 P.M.). The price was so good that it was $1,200 less than the same dealership quoted to me for the same car on the Internet. How do I know it was the same car—I mean really the same car? The price was so good that I asked the salesperson for the VIN (that is, the vehicle identification number), and he gave it to me.

Why does this method work? Game theory is about strategizing in your dealings with other people. Part of strategizing involves realizing that the other person is doing the same. Pat, the salesperson, is thinking about what to say to get a better price from you. You have to think through different ways to respond to Pat's arguments. The central concern is to anticipate how Pat is likely to reply to you if you say a really high price, a really low price, a moderate price, or, as I recommend, no price at all. Let Pat put the price on the table knowing that lots of other dealers will be given the same opportunity and that you have no intention of keeping Pat's offer a secret. When Pat is asked for the dealership's best price, Pat knows that this is the one and only chance they will have to sell you a car.

Salespeople want you to feel good about them. They want you to feel you are letting them down if you hold out for that last fifty dollars or if you question their sincerity. They believe this will weaken your resolve. Please remember: These are the people who invented "dealer prep"! They just want the best price they can get. They have no plans to be your friend.

The telephone approach solves all of these problems. On the telephone, body language is eliminated from the equation and you, not the seller, are in control of the conversation.[2] You have set the sequence of moves and defined the game. They know that you will speak to enough dealers that anyone who is on the cusp of a manufacturer's incentive, for instance, will give you a really good price. They reveal information to you however they reply. The fact that some of them leave the conversation early is a good thing—it saves you time. You have not lost an opportunity. They know their market; that is part of their business. You have given them an incentive to be truthful. They are better off for it, and most assuredly so are you.

When you buy a car this way, that day at 5:00 P.M. you make a deal that is as close to the seller's minimum price as you can ever hope to get. You are in charge of the negotiation and entitled to feel really good about the deal you struck.

You may be surprised to learn that buying cars and negotiating international crises are not all that different. In fact, I'll show you how we go from buying a car to negotiating the North Korean nuclear threat. But before we do that, it's time to dissect game theory a bit to reveal the principles that allow us to anticipate and shape the future.

2

■

GAME THEORY 101

WHEN PEOPLE TALK about science, subjects like chemistry and physics leap to mind. Political science certainly does not. But science is a method, not a subject. It is a method that relies on logical arguments and experimental evidence to figure out how the world of things—and of people—works. The scientific method certainly applies to politics just as it does to physics. Still, physics and politics are quite obviously entirely different subjects. One of the ways they differ is crucial for understanding everything that is to come. You see, the world of physics is pretty much about how particles interact. Now, the central feature of particle interactions is that photons, electrons, neutrons, or their constituent quarks never anticipate crashing into one another. Consequently, there is no strategizing behind the collision of particles.

Studying people is ever so much more complicated than studying inanimate particles. Just think how different interactions are between quarks and Quakers, electrons and electors, protons and protesters. People, and in fact just about every living thing, seem to have a survival instinct. Genes act as if they want to get passed on, bacteria find hosts, cockroaches flee my shoe, and ordinary people look out for what they think is good for them and try to avoid what they think is bad for them. That includes cooperating with friends and fighting with foes. Like the physicist's particles, people interact, but unlike the physicist's particles, people interact strategically. That is what game-theory thinking is all about.

To be a successful prognosticator, it is critical to think about how other people think about their problems. It is just as important to think about how other people think about how you think about your problems and theirs. The previous, tedious sentence, by the way, could be repeated ad infinitum to reflect on the information that gets ferreted out when thinking strategically. This and the next chapter—and the science of predictioneering—are about solving the problem of working out what others think, what they think you think, what you think they think, what you think they think you think. . . . This is the kind of information that physicists rightly don't give a moment's thought to when studying the particles that capture their interest—but it is the foundation from which we can see when and how to turn situations to our own advantage.

WHERE WE ARE HEADED

In Game Theory 101 we'll consider how to look at the world through the eyes of others. For starters, we'll need to set aside, at least for argument's sake, our natural optimism about human nature. Game theory urges us to take a cold, hard look at what it means to be a calculating, rational decision maker. Sure, there are some genuinely nice, altruistic people in the world—but that doesn't mean they aren't carefully calculating their actions. In fact, we'll see that even as nice and altruistic a person as Mother Teresa can be scrutinized through the not-so-warm-and-fuzzy eyes of a game theorist. Doing so will help us understand how paths as different as hers and a suicide bomber's can be equally rational and strategically sensible. It will also help us realize that even some of the most unquestioned received wisdom—such as the existence of something called the national interest—may be just a strategic fiction created by politicians for their own advantage instead of ours. Depressing? Yes. Accurate? You bet.

This chapter will provide us with a framework for the game theorist's notions of interests, beliefs, and rationality; a sense of how to use logic to cut through the fog of language; and an understanding of strategic behavior that, in conjunction with the previous two concepts, leads to an ability to better map and anticipate the thinking and actions of others.

WHAT ARE THE OTHER GUY'S INTERESTS AND BELIEFS?

Game theory comes in two primary flavors. *Cooperative game theory* was invented by John von Neumann and Oskar Morgenstern.[1] Their 1947 book on the subject drew a clear and compelling analogy between problems people (or nations) face and parlor games like charades or the name-in-the-hat game, a favorite in my family. These sorts of games deal with players who engage each other, trying to anticipate moves and countermoves, but only in a setting where what they say they will do is the same as what they actually do. That's why it's called cooperative game theory—a promise made is a promise kept. Because of this, one big limitation with cooperative game theory, especially in games that involve more than two players, is that it has far too optimistic a view of human nature. In this universe people make deals and keep them. They can be bought off, sure, but once they say they'll do something, they do it. That means cooperative game theory works fine for zero-sum games where what one side loses equals what the other side wins, but not all that many interesting problems in the world are that cut and dried. When they are not, this original variety of game theory is not nearly as good for my purposes as what has replaced it.

By the early 1950s, the mathematician John Nash, the subject of *A Beautiful Mind* and the winner of the 1994 Nobel Prize in Economics, invented a different kind of game theory.[2] He drew attention to the propensity people have *not* to cooperate with one another. Poker players and diplomats use polite terms, like "bluffing," for what ordinary people mean when they say someone is a liar. In noncooperative games, promises do not necessarily mean anything. Lies are a part of strategizing. Promises are kept when a player decides it's in her interest to do what she promised. When promises and interests differ, people renege, they break their word, they cheat, they do whatever they think will benefit them most. Of course, they know that bluffing and cheating can be costly. Therefore, they take prospective costs as well as benefits into account. In fact, raising costs is one way, albeit a difficult and painful way, of encouraging people to be truthful. Indeed, that is exactly the purpose behind meeting and then raising someone's bet in poker or calling car dealers instead of going in to see them.

The view of people as cold, ruthless, and self-interested is at the heart of game-theory thinking. There may be room for nice guys, but not much. Most of the time, nice guys really do finish last. Those who will throw themselves on a hand grenade to save their fellows, well, they do so and then, tragically, they are dead. They are out of the game of life. We remember them, we honor them, we extol them, but we just don't compete with them, because they are not here to compete with us. Such good souls need not occupy much of our time. Or, if they do, we applied game theorists take a cynical view and look for how suicide might benefit them. There might be virgins in heaven, or, as with kamikaze pilots in World War II, Crusaders in the Middle Ages, and some suicide bombers today, there might be significant financial incentives such as cash payments and debt forgiveness to their families in exchange for their sacrifice.

Some may find this materialistic explanation of personal sacrifice offensive. The trouble is, it's a lot costlier to believe mistakenly in other people's goodwill than it is to be a cynic and assume they're looking out for themselves (until and unless their actions say otherwise). It is hard to get burned in personal dealings if you remember Ronald Reagan's dictum: Trust but verify. For those who are offended by this tough view of human nature, I urge you to consider some facts.

The United States operates the Concerned Local Citizens program in Iraq. Following the alphabet-soup tradition so beloved by the Pentagon, the Iraqis participating in this program are known as CLCs. CLCs help guard neighborhoods against insurgents. They are paid ten dollars a day for their service. It doesn't seem as if there is anything crass or overly materialistic about that. But then we should pause to ask, who are these CLCs and what, exactly, are we buying for ten dollars a day?

These concerned Iraqis are not your ordinary neighborhood watch group. They are not the folks next door who give school kids a safe place to go when their parents are at work. They are not the friends who have your house key, water your plants, take in your mail, and feed your cat while you're on vacation. No, they're former anti-American insurgents, tens of thousands of them. Some of them, in fact, used to belong to al-Qaeda. It would seem that they were among the most fanatic of fanatics, the worst of the worst. And yet for a measly ten bucks a day these supposedly unshakable al-Qaeda terrorists now act like allies of the United States, serving as our very own paramilitaries, helping to keep violence down in mostly Sunni neighborhoods, defending the peace that they used to shatter for a living. How can this be? How can terrorists be so easily converted into our friends and protectors?

As it happens, being an ex-insurgent employed as a CLC is a very good job by Iraqi standards. At ten dollars a day, CLCs can earn a few thousand dollars a year from the United States, plus, of course, whatever extra they make on the side. The average Iraqi, despite that country's huge oil wealth, earns only about six dollars a day, almost half what a CLC gets![3] Those who think that terrorists are irrational religious zealots who do not respond to monetary and personal incentives should remember that a daily dose of just ten dollars is enough to get such folks to become quasi-friends of the United States of America.

Of course, there is as much room for saints as for sinners in game theory. There's no problem accommodating the (few) Mother Teresas of our world. Since game theory is about choosing actions given expected costs and benefits, it does encourage us to ask, perhaps obnoxiously, what benefits Mother Teresa might have expected in return for her life of sacrifice and good works. We cannot help but notice that she did not serve the poor as quietly as most nuns do, living out their lives in anonymous obscurity. The very publicness of Mother Teresa's deeds reassures us of her rationality and her potential to help poor people on a large scale.

Whether we call on the Catholic understanding of a saintly life or the Talmudic view of a charitable life, we encounter a problem on Mother Teresa's behalf. In doing her good works, she might have had to worry, as (Saint) Bernard of Clairveaux (1090–1153) did, that in obeying God's commandments as faithfully as possible she could be committing the deadly sin of pride. Maybe she thought herself better than others, more deserving of heaven, even worthy of sainthood, exactly because of her personal sacrifice and good works. That, as we will see, does not seem to have been a major source of worry for her.

From the Talmudic perspective as expressed by Moses Maimonides (1135–1204), she would have had at least as big a problem. Maimonides, or Rambam as he was known in his day, concluded that charity given anonymously to anonymous recipients in order to help them become self-sufficient is the best kind. Mother Teresa's giving did not rise to this standard, and she made sure it didn't. She did not give anonymously; she knew to whom she was giving; and she did not strive particularly to make the beneficiaries of her kindness self-sufficient. In fact, she went out of her way to make herself and her acts recognizable. For instance, Mother Teresa carefully promoted herself, creating brand-name recognition—just like Cheerios, Coke, Xerox, or Vaseline—by always wearing the special habit of the order she founded (a white sari with blue trim and sandals) so that she could not be easily confused with just any nice old lady. Of course, anonymous giving could still be prideful, but for sure it could not lead to a Nobel Peace Prize in this world or to beatification and canonization in the next.

Could it be that Mother Teresa's ambition for herself was tied to her faith in an eternal reward? It makes sense to pay the price of sacrifice for the short, finite time of a life span if the consequence is a reward that goes on for infinity in heaven. In fact, isn't that exactly the explanation many of us give for the actions of suicide bombers, dying in their own prideful eyes as martyrs who will be rewarded for all eternity in heaven?

Or maybe, in Mother Teresa's case, the rational, calculating motivation behind her deeds was more complex. We know now that she questioned her religious faith and the existence of God.[4] Her doubts apparently began shortly after she started to minister to the poor and sick in Calcutta. By then maybe she felt locked into the religious life she chose for herself. Doubting God and ill-prepared for a life outside the Church, perhaps she found a perfect strategy for gaining the acclaim in life that she feared might not exist after death. Was she looking for an eternal reward, or for reward in the here and now? Only she could really know. We applied game theorists are content to observe that she acted *as if* being rewarded was her motivation. That is, she was not cold and materialistic; she was warm and materialistic. That is enough to make her a fine subject for analysis as a rational, strategic player in the game of life—and maybe enough to earn her sainthood as well.

Game theory draws our attention to important principles that shape what people say and do. First of all, just like Mother Teresa or a suicide bomber, all people are taken to be rational. That just means we assume they do what they believe is in their own best interest, whether that's making as much money as they can or gaining entry to heaven or anything else. They may find out later that they made a poor choice, but in game-theory thinking we worry about what people know, believe, and value at the time they choose their actions, not what they find out later when it's too late to do something else. Game theory has no place for Monday-morning quarterbacks. It's all about what to do when decisions must be made, even if we cannot know for sure what the consequences of our actions will be.

This notion of rational action seems to trouble some people. Usually that's because they mean something different from what an economist or political scientist means when talking about rationality. Words can have many meanings, so we must be careful to define ideas carefully. As it happens, game theorists insist on a particular use of the word "rational."

Some folks seem to think that rational people must be super smart, never making a mistake, looking over each and every possible thing that could happen to them, working out the exact costs and benefits of every conceivable course of action. That is nonsense. Nobody is that smart or diligent, nor should they be. Actually, checking out every possible course of action, working out everything that possibly could arise, is almost never rational, at least not as the term is used in my world. It is never rational to continue searching for more information, for example, when the cost of finding out more is greater than the expected benefits of knowing more. Rational people know when to stop searching—when enough is enough. (I try to impart this message to my students. When they tell me they want to make their term papers as good as possible, I plead with them not to. A paper that is worked on until it is as good as possible will never be finished.)

Another way that people talk about rationality that has nothing to do with what "rational choice theorists" have in mind is to discuss whether what someone wants is rational or not. Distasteful as the fact may be, people with crazy ideas can be perfectly rational. Rationality is about choosing actions that are consistent with advancing personal interests, whatever those interests may be. It has nothing to do with whether you or I think what someone wants is a good idea, shows good taste or judgment, or even makes sense to want.

I certainly think what Adolf Hitler said he wanted and what he did to advance his heinous goals were evil, but I am reluctant to let him off the hook with an insanity plea by saying he was not rational. His actions were rational given his evil aims, and therefore it was perfectly right and proper to hold him and his henchmen accountable.

The same holds for modern-day terrorists. They're not nuts. They are desperate, calculating, disgruntled people who are looking for ways to force others to pay attention to their real or perceived woes. Dismissing them as irrational misses the point and leads us to make wrongheaded choices about how to handle their threat. We do ourselves no service by labeling people as insane or irrational simply because we can't understand their goals. Our attention is better fixed on what they do, since we probably can change or impede their actions even when we can't alter what they want.

What exactly does rationality require? Actually it's a simple idea. To be rational, a person must be able to state a preference among choices, including having no preference at all (that is, being truly indifferent). Also, their preferences must not go in circles. For instance, if I like chocolate ice cream better than vanilla—who doesn't?—and vanilla better than strawberry, then I also presumably like chocolate ice cream better than strawberry. Finally, rational people act in accordance with their preferences, taking into account the impediments to doing so. For instance, one ice cream parlor might be sold out of chocolate more often than another. I might be willing to risk having to settle for vanilla if the place that runs out also has much better tasting chocolate. Taking calculated risks is part of being rational. I just need to think about the size of the risk, the value of the reward that comes with success, and the cost that comes with failure, and compare those to the risks, costs, and benefits of doing things differently.

Since rational people take calculated risks, sometimes things turn out badly for them. Nobody gets everything they want. I sometimes end up drinking soda I don't like or eating vanilla or strawberry ice cream despite my best efforts to obtain what I prefer. That's what it means to take risks. We absolutely cannot conclude that someone was irrational or acted irrationally just because at the end of the day they got a rotten outcome, whether that means being stuck with strawberry ice cream, losing a war, or even worse.

Rational choices reflect not only thinking through risks but also trying to sort out costs and benefits. Costs and benefits can be tricky to work out. I could be unsure of what those costs or benefits are likely to be. That too can be an important impediment or constraint on my rational decisions. Sometimes we have to make decisions even though we are in the dark about the consequences. Fortunately, that doesn't happen much with buying ice cream or soda, but it sure happens a lot when negotiating a big business deal or forging a new foreign policy. In those cases, we had better be careful to weigh the sources of our uncertainty carefully, and not plunge headlong into some dangerous endeavor with no more than rose-colored glasses to guide our way. We may not get the consequence we want, but we can be careful to manage the range of consequences that are likely to arise. (Just imagine how different the debacle in Iraq might have been, for example, had American leaders not thought that the Iraqi people would be dancing in the streets, kissing American soldiers after Saddam was overthrown the way Parisians did when Americans marched into Paris behind Charles de Gaulle on August 26, 1944.)

The question remains, however, as to when someone is actually irrational. In everyday usage, lots of behavior looks irrational even though on closer inspection it turns out not to be. Sometimes critics point to behavior like leaving tips in restaurants, giving gifts to friends, or—sorry, I don't mean to be gross—flushing the toilet in public places like airports or museums as irrational acts. They argue that all of the benefit goes to someone else, not to the tipper, gift giver, or flusher. I say, not true.

Many rational acts impose short-term costs on the doer with the expectation of longer-term gains. That's true of tipping, gift giving, flushing public toilets, not littering, and lots more. Sure, you might leave a tip even though you don't expect to be in the particular restaurant again. Tipping, however, like gift giving, is a social norm that has arisen and taken hold because we have learned that its effects on the expectations of others (waiters, dinner party hosts) are important to making our own lives a little happier and easier. If waiters thought they weren't going to get a tip and yet continued to be paid poorly, then it's a good bet that service would be much worse in every restaurant. Studies show, for instance, that customer satisfaction with service does not help predict the restaurants people choose in southern China.[5] Tipping is illegal in China (which is not to say that it never happens, but it isn't expected). It is good to keep in mind that people act on expectations. It seems that the quality of service doesn't vary much between restaurants in southern China, because the service ethic just isn't guided by anticipated rewards for good service. Take away the expectation of tips, and the waitstaff is motivated by something other than the customers' interests and the waiters' rewards for satisfying those interests.

Tipping, gift giving, and, yes, flushing the toilet create good expectations that make each of us better off most of the time even if they cost a little at the moment. Sure, we could free-ride on the good acts of others, save a little money or the little bit of effort it takes to flush a toilet or throw litter in the garbage can instead of on the street, but most of us would feel bad about ourselves if we did that. The urge to feel good about ourselves—not to take the risk of offending others and not to bear the cost of their reaction—is sufficient to induce us to behave in a socially appropriate way. For the few misanthropes who prefer to save the money that a tip or a gift costs or the effort that flushing a toilet costs, well, they are behaving rationally too. They aren't concerned about feeling like lowlifes. They value the savings from their poor behavior more than goodwill or long-term good results. That's why there really is no accounting for taste. Rationality is, as I said, about doing what you believe is in your own interest; it doesn't impose interests on us.

So what *does* constitute irrationality in an applied game theorist's world? A person is irrational if, returning to the example of ice cream flavors, all of the following are true: she likes strawberry ice cream better than chocolate; strawberry ice cream costs no more than chocolate ice cream; strawberry ice cream is readily available for purchase; and still she goes and buys chocolate ice cream for herself. In such a case, I might wonder whether she had eaten so much strawberry ice cream recently that she wanted a change (a preference for variety over constancy, adding another dimension to the things preferred that was not included on my list) or something like that, but if those sorts of considerations are absent, then a strawberry lover is expected to eat strawberry ice cream when everything else is equal.

All of this is to say that, really, the only people who are ruled out by assuming rationality are very little children and perhaps schizophrenics. Little children—most especially two-year-olds—and schizophrenics sometimes act as if their preferences change every few seconds. One minute they want strawberry and the next it's the worst thing in the world. That sort of flip-flopping in individual preferences is hazardous for those who want to predict or engineer people's choices. Reasoning with people who flip-flop all the time is all but impossible. They're not committed to being logically consistent in what they say, want, or do.

Nature may not abhor a vacuum, but game theory definitely abhors logical inconsistency. If you allow the possibility that what an individual *really* wants changes all the time, moment to moment, then you can claim that anything they do and anything they get fits in with (or contradicts) their interests. That certainly won't lead to good predictions or good engineering, and besides, it just isn't any fun. It takes all of the challenge out of working out what people are likely to do.

WHAT IS THE OTHER GUY'S LOGIC (NOT HIS LANGUAGE)?

On account of the above, it may have become readily apparent to you that game theory alerts us to be careful in how we express and understand our interests and those of others. It's easy to make logical mistakes, and they can be hard to spot, which can often disguise or obscure the meaning of the thinking and actions of individuals. That is why game theorists use mathematics to work out what people are likely to do.

Ordinary everyday language can be awfully vague and ambiguous. A friend of mine is a linguist. One of his favorite sentences goes like this: "I saw the man with a telescope." Now that is one vague sentence. Did I look through a telescope and spot a man, or did I look over at a man who was carrying a telescope, or does the sentence mean something entirely different? You can see why linguists like this sentence. It gives them an interesting problem to work out. I don't like sentences like that. I like sentences written with mathematics (and so do many linguists). They don't produce poetic beauty or double entendres, which makes them boring, but it also gives them a great virtue. In English, saying things are equal often means "more or less"; in math, "equal" means just that, equal, not almost equal or usually equal, but plain simple equal.

We humans have devised all sorts of clever ways to cover up sloppy or slippery arguments. As I am fond of telling my students, my suspicions are aroused by sentences beginning with clauses like "It stands to reason that" or "It is a fact that. . . ." Usually, what follows the statement "It stands to reason that" does not. The clause is being asked to substitute for the hard work of showing that a conclusion follows logically from the assumptions. Likewise, "It is a fact that" generally precedes an expression of opinion rather than a fact. Watch out for these. This sort of rhetoric can easily take a person down a wrong line of thinking by accepting as true some-thing that might be true and then again might not be.

Consider, for example, what policies you think our national leaders should follow to protect and enhance our national interest. When we think carefully about how to further the national interest, it becomes evi-dent that sometimes things that seem obviously true are not, and that a little logic can go a long way to clarify our understanding.

It is commonplace to think that foreign policy should advance the na-tional interest. This idea is so widespread that we accept it as an obvious truth, but is it? We hardly ever pause to ask how we know what is in the national interest. Most of the time, we seem to mean that policies bene-fiting the great majority of people are policies in the national interest. Se-cure borders to prevent foreign invasions or illegal immigration are thought to be in the national interest. Economic policies that make citi-zens more prosperous are thought to be in the national interest. Yet we also know that money spent on defending our national security is money that is not spent on building the economy. There is a trade-off between the two. What, then, is the right balance between national security and economic security that ensures the national interest?

Imagine that American citizens are divided into three equally sized groups. One group wants to spend more on national defense and to adopt more free-trade programs. Call these people Republicans. Another wants to cut defense spending and shift trade policy away from the status quo in order to better protect American industry against foreign competition. Call them Democrats. A third wants to spend more on national defense and also to greatly increase tariffs to keep our markets from being flooded with cheap foreign-made goods. Call this faction blue-collar indepen-dents. With all of these voters in mind, what defense and trade policy can rightfully call itself "the national interest"? The answer, as seen in figure 2.1 (on the next page), is that *any* policy can legitimately lay claim to being in—or against—the national interest.

Figure 2.1 places each of our three voting blocs—Republicans, Dem-ocrats, and blue-collar independents—at the policy outcomes they prefer when it comes to trade and defense spending. That's why Republicans are found in the upper right-hand corner as you look at the figure, indicating their support for much freer trade and much higher defense spending.

FIG. 2.1. Defense and Trade Policy in the National Interest

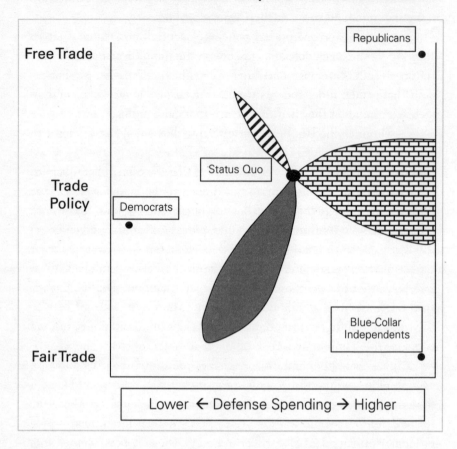

Democrats are on the far left-hand side just below the vertical center. That is consistent with their wanting much less spent on defense and a modest shift in trade policy. Blue-collar independents are found on the bottom right, consistent with their preference for trade protection and higher defense outlays. And, as you can see, there is a point labeled "Status Quo," which denotes current defense spending and trade policy.

By putting the two issues together in the figure I am acknowledging that they are often linked in public debate. The debate generally revolves around how best to balance trade and defense given that there are inherent trade-offs between them. Free trade, for instance, can imply selling high-end computer technology, weapons technology, and other technologies that adversaries might use to threaten our national security. High tariffs might provoke trade wars or worse, thereby potentially harming national security and prompting arguments to spend more on national defense.

I assume that everyone prefers policies closer to their favored position (that's where the black dots associated with the Republicans, Democrats, and independents are positioned) to policies that are farther away. For example, blue-collar independents would vote to change the status quo on defense and trade if they had the chance to choose a mix on these issues that was closer to the black dot associated with them—that is, closer to what they want.

To show the range of policy combinations that the blue-collar independents like better than the status quo, I drew a circle (showing only a part of it) whose center is their most desired policy combination and whose perimeter just passes through the status quo policy.[6] Anything inside the arc whose center is what blue-collar independents most want is better for them than the prevailing approach to defense spending and trade. The same is true for the points inside the arcs centered on the Republicans and the Democrats that pass through the status quo.

By drawing these circles around each player's preferred policy mix we learn something important. We see that these circles overlap. The areas of overlap show us policy combinations that improve on the status quo for a coalition of two of the three players. For instance, the lined oblong area tilting toward the upper left of the figure depicts policies that improve the well-being of Democrats and Republicans (ah, a bipartisan foreign policy opposed by independent blue-collar workers). The gray petal-shaped area improves the interests of Democrats and blue-collar independents (at the expense of Republicans), and the bricked-over area provides a mix of trade and defense spending that benefit the Republicans and blue-collar independents (to the chagrin of Democrats).

Because we assumed that each of the three voting blocs is equal in size, each overlapping area identifies defense and trade policies that command the support of two-thirds of the electorate. Here's the rub, then, when it comes to talking about the national interest. One coalition wants more free trade and less defense spending. Another wants less free trade and less defense spending. The third wants less free trade and more defense spending. So, we can assemble a two-thirds majority for more defense spending and also for less. We can find a two-thirds coalition for more free trade or for higher tariffs or (in the politically charged rhetoric of trade debate) for more fair trade. In fact, there are loads of ways to allocate effort between defense spending and trade policy to make better off whichever coalition forms.[7]

What, then, is the national interest? We might have to conclude that except under the direst circumstances there is no such thing as "the national interest," even if the term refers to what a large majority favors. That is surprising, perhaps, but it follows logically from the idea that people will align themselves behind policies that are closer to what they want against policies that are farther from what they advocate. It just happens that any time there are trade-offs between alternative ways to spend money or to exert influence, there are likely to be many different spending or influence combinations that beat the prevailing view. None can be said to be a truer reflection of the national interest than another; that reflection is in the eyes of the beholder, not in some objective assessment of national well-being. So much for the venerable notion that our leaders pursue the national interest, or, for that matter, that business executives single-mindedly foster shareholder value. I suppose, freed as they are to build a coalition that wants whatever it is they also want, that our leaders really are free to pursue their *own* interests and to call that the national interest or the corporate interest.

WHAT IS THE OTHER GUY'S BEHAVIOR? (DOES HE HAVE GOOD CARDS OR NOT?)

To understand how interests frame so many of the questions we have at stake, game theory still requires that people behave in a logically consistent way within those interests. That does not mean that people cannot behave in surprising ways, for surely they can. If you've ever played the game Mastermind, you've confronted the difficulties of logic directly. In Mastermind—a game I've used with students to teach them about really probing their beliefs—one player sets up four (or, in harder versions, more) colored pegs selected from among six colors in whatever order he or she chooses. The rest of the players cannot see the pegs. They propose color sequences of pegs and are told that yes, they got three colors right, or no, they didn't get any right, or yes, they got one color in the right position but none of the others. In this way, information accumulates from round to round. By keeping careful track of the information about what is true and what is false, you gradually eliminate hypotheses and converge on a correct view of what order the colored pegs are in. This is the point behind a game like Mastermind, Battleship, or charades. It is also one point behind the forecasting games I designed and use to predict and engineer events.

The key to any of these games is sorting out the difference between knowledge and beliefs. Different players in any game are likely to start out with different beliefs because they don't have enough information to know the true lay of the land. It is fine to sustain beliefs that could be consistent with what's observed, but it's not sensible to hold on to beliefs after they have been refuted by what is happening around us. Of course, sorting out when beliefs and actions are inconsistent requires working out the incentives people have to lie, mislead, bluff, and cheat.

In Mastermind this is easy to do because the game has rules that stipulate the order of guessing and that require the person who placed the pegs to respond honestly to proposed color sequences suggested by other players. There is no point to the game if the person placing the pegs lies to everyone else. But even when everyone tells the truth, it is easy to slip into serious lapses in logic that can lead to entirely wrong beliefs. That is something to be careful about.

Slipping into wrong beliefs is a problem for many of us. It is easy to look at facts selectively and to reach wrong conclusions. That is a major problem, for instance, with the alleged police practice of profiling, or some people's judgment about the guilt or innocence of others based on thin evidence that is wrongly assessed. There are very good reasons why the police and we ordinary folk ought not to be too hasty in jumping to conclusions.

Let me give an example to help flesh out how easily we can slip into poor logical thinking. Baseball is beset by a scandal over performance-enhancing drugs. Suppose you know that the odds someone will test positive for steroids are 90 percent if they actually used steroids. Does that mean when someone tests positive we can be very confident that they used steroids? Journalists seem to think so. Congress seems to think so. But it just isn't so. To formulate a policy we need an answer to the question, How likely is it that someone used steroids if they test positive? It is not enough to know how likely they are to test positive if they use steroids.

Unfortunately, we cannot easily know the answer to the question we really care about. We can know whether someone tested positive, but that could be a terrible basis for deciding whether the person cheated. A logically consistent use of probabilities—working out the real risks—can help make that clear.

Imagine that out of every 100 baseball players (only) 10 cheat by taking steroids (game theory notwithstanding, I am an optimist) and that the tests are accurate enough that 9 out of every 10 cheaters test positive. To evaluate the likelihood of guilt or innocence we still need to know how many honest players test positive—that is, how often the tests give us a false positive answer. Tests are, after all, far from perfect. Just imagine that while 90 out of every 100 players do not cheat, 10 percent of the honest players nevertheless test (falsely) positive. Looking at these numbers it's easy to think, well, hardly anyone gets a false positive (only 10 percent of the innocent) and almost every guilty party gets a true positive (90 percent of the guilty), so knowing whether a person tested positive must make us very confident of their guilt. Wrong![8]

With the numbers just laid out, 9 out of 10 cheaters test positive and 9 out of 90 innocent ball players also test positive. So, 9 out of 18 of the positive test results include cheaters and 9 out of 18 include absolutely innocent baseball players. In this example, the odds that a player testing positive actually uses steroids are fifty-fifty, just the flip of a coin. That is hardly enough to ruin a person's career and reputation. Who would want to convict so many innocents just to get the guilty few? It is best to take seriously the dictum "innocent until proven guilty."

The calculation we just did is an example of Bayes' Theorem.[9] It provides a logically sound way to avoid inconsistencies between what we thought was true (a positive test means a player uses steroids) and new information that comes our way (half of all players testing positive do not use steroids). Bayes' Theorem compels us to ask probing questions about what we observe. Instead of asking, "What are the odds that a baseball player uses performance-enhancing drugs?" we ask, "What are the odds that a baseball player uses performance-enhancing drugs given that we know he tested positive for such drugs and we know the odds of testing positive under different conditions?"

Bayes' Theorem provides a way to calculate how people digest new information. It assumes that everyone uses such information to check whether what they believe is consistent with their new knowledge. It highlights how our beliefs change—how they are updated, in game-theory jargon—in response to new information that reinforces or contradicts what we thought was true. In that way, the theorem, and the game theorists who rely on it, view beliefs as malleable rather than as unalterable biases lurking in a person's head.

This idea of updating beliefs leads us to the next challenge. Suppose a baseball player who had a positive (guilty) test result is called to testify before Congress in the steroid scandal. Now suppose he knows of the odds sketched above. Aware of these statistics, and knowing that any self-respecting congressperson is also aware of them, the baseball player knows that Congress, if citing only a positive test result as their evidence, in fact has little on him, no matter how much outrage they muster. The player, in other words, knows Congress is bluffing. But of course Congress knows this as well, so they have subpoenaed the player's trainer, who is coming in to testify right after the player. Is this just another bluff by Congress, tightening the screws to elicit a confession with the threat of perjury looming? Whether the player is guilty or not, perhaps he shrugs off the move, in effect calling Congress's raising of the stakes. Now what? Does Congress actually have anything, or will they be embarrassed for going on a fishing expedition and dragging an apparently innocent man through the mud? Will the player adamantly profess innocence knowing he's guilty (but maybe he really isn't), and should we shrug off the declarations of innocence lightly, as it seems so many of us do? Is Congress bluffing? Is the player bluffing? Is everyone bluffing? These are tough problems, and they are right up game theory's alley!

In real life there are plenty of incentives for others (and for us) to lie. That is certainly true for athletes, corporate executives, national leaders, poker players, and all the rest of us. Therefore, to predict the future we have to reflect on when people are likely to lie and when they are most likely to tell the truth. In engineering the future, our task is to find the right incentives so that people tell the truth, or so that, when it helps our cause, they believe our lies.

One way of eliciting honest responses is to make repeated lying really costly. Bluffing at poker, for instance, can be costly exactly because other players sometimes don't believe big bets, and don't fold as a result. If their hand is better, the bluff comes straight out of the liar's pocket. So the central feature of a game like five-card draw is not figuring out the probability of drawing an inside straight or three of a kind, although that's certainly useful too. It's about convincing others that your hand is stronger than it really is. Part of the key to accumulating bargaining chips, whether in poker or diplomacy, is engineering the future by exploiting leverage that really does not exist. Along with taking prudent risks, creating leverage is one of the most important features in changing outcomes. Of course, that is just a polite way of saying that it's good to know when and how to lie.

Betting, whether with chips, stockholders' money, perjury charges, or soldiers, can lead others to wrong inferences that benefit the bettor; but gambling always suffers from two limitations. First, it can be expensive to bet more than a hand is worth. Second, everyone has an interest in trying to figure out who is bluffing and who is being honest. Raising the stakes helps flush out the bluffers. The bigger the cumulative bet, the costlier it is to pretend to have the resolve to see a dispute through when the cards really are lousy. How much pain anyone is willing to risk on a bluff, and how similar their wagering is when they are bluffing and when they are really holding good cards, is crucial to the prospects of winning or of being unmasked. That, of course, is why diplomats, lawyers, and poker players need a good poker face, and it is why, for example, you take your broker's advice more seriously if she invests a lot of her own money in a stock she's recommending.

Getting the best results comes down to matching actions to beliefs. Gradually, under the right circumstances, exploiting information leads to consistency between what people see, what they think, and what they do, just as it does in Mastermind. Convergence in thinking facilitates deals, bargains, and the resolution of disputes.

With that, we've just completed the introductory course in game theory. Nicely done! Now we're ready to go on to the more advanced course. In the next chapter we look in more depth at how the very fact of our being strategic changes everything going on around us. That will set the stage for working out how we can use strategy to change things to be better for ourselves and those we care about and, if we are altruistic enough, maybe even for almost everyone.

3

■

GAME THEORY 102

AME THEORY 101 started us off thinking about how different people are from particles. In short, we are strategists. We calculate before we interact. And with 101 under our belts, we know enough to delve more closely into the subtleties of strategizing.

Of the many lessons game theory teaches us, one of particular import is that the future—or at least its anticipation—can cause the past, perhaps even more often than the past causes the future. Sound crazy? Ask yourself, do Christmas tree sales cause Christmas? This sort of reverse causality is fundamental to how game theorists work through problems to anticipate outcomes. It is very different from conventional linear thinking. Let me offer an example where failing to recognize how the future shapes the past can lead to really bad consequences.

Many believe that arms races cause war.[1] With that conviction in mind, policy makers vigorously pursue arms control agreements to improve the prospects of peace. To be sure, controlling arms means that if there is war, fewer people are killed and less property is destroyed. That is certainly a good thing, but that is not why people advocate arms control. They want to make war less likely. But reducing the amount or lethality of arms just does not do that.

The standard account of how arms races cause war involves what game theorists call a hand wave—that is, at some point the analyst waves his hands in the air instead of providing the logical connection from argument to conclusions. The arms-race hand wave goes like this:

When a country builds up its arms it makes its adversaries fear that their security is at risk. In response, they build up their own arms to defend themselves. The other side looks at that buildup—seeing their own as purely defensive—and tries to protect itself by developing still more and better weapons. Eventually the arms race produces a massive overcapacity to kill and destroy. Remember how many times over the U.S. and Soviet nuclear arsenals could destroy the world! So, as the level of arms—ten thousand nuclear-tipped missiles, for instance—grows out of proportion to the threat, things spiral out of control (that's the hand wave—*why do things spiral out of control?*), and war starts.

Wait a moment, let's slow down and think about that. The argument boils down to claiming that when the costs of war get to be really big—arms are out of proportion to the threat—war becomes more likely. That's really odd. Common sense and basic economics teach us that when the cost of anything goes up, we generally buy less, not more. Why should that be any less true of war?

True, just about every war has been preceded by a buildup in weapons, but that is not the relevant observation. It is akin to looking at a baseball player's positive test for steroids as proof that he cheats. What we want to know is how often the acquisition of lots more weapons leads to war, not how often wars are preceded by the purchase of arms. The answer to the question we care about is, *not very often.*

By looking at wars and then asking whether there had been an arms race, we confuse cause and effect. We ignore all the instances in which arms may successfully deter fighting exactly because the anticipated destruction is so high. Big wars are very rare precisely because when we expect high costs we look for ways to compromise. That, for instance, is why the 1962 Cuban Missile Crisis ended peacefully. That is why every major crisis between the United States and the Soviet Union throughout the cold war ended without the initiation of a hot war. The fear of nuclear annihilation kept it cold. That is why lots of events that could have ignited world wars ended peacefully and are now all but forgotten.

So, in war and especially in peace, reverse causality is at work. When policy makers turn to arms control deals, thinking they are promoting peace, they are taking much bigger risks than they seem to realize. Failing to think about reverse causation leads to poor predictions of what is likely to happen, and that can lead to dangerous decisions and even to cata-strophic war.

We will see many more instances of this kind of reasoning in later chapters. We will examine, for example, why most corporate fraud proba-bly is not sparked by executive greed and why treaties to control green-house gas emissions may not be the best way to fight global warning. Each example reinforces the idea that correlation is not causation. They also re-mind us that the logic of reverse causation—called endogeneity in game theory—means that what we actually "observe"—such as arms races fol-lowed by war—are often biased samples.

The fact that decisions can be altered by the expectation of their con-sequences has lots of implications. In Game Theory 101 we talked about bluffing. Working out when promises or threats should be taken seriously and when they are (in game-theory-speak) "cheap talk" is fundamental to solving complicated situations in business, in politics, and in our daily en-counters. Sorting out when promises or threats are sincere and when they are just talk is the problem of determining whether commitments are credible.

LET'S PLAY GAMES

In predicting and engineering the future, part of getting things right is working out what stands in the way of this or that particular outcome. Even after pots of money are won at cards, or hands are shaken and con-tracts or treaties are signed, we can't be sure of what will actually get implemented. We always have to ask about commitments. Deals and promises, however sincerely made, can unravel for lots of reasons. Econ-omists have come up with a superbly descriptive label for a problem in en-forcing contracts. They ask, is the contract "renegotiation-proof"?[3] This question is at the heart of litigiousness in the United States.

I once worked on a lawsuit involving two power companies. One pro-duced excess electricity and sold it to a different electric company in an-other state. As it happened, the price for electricity shot way up after the contract was signed. The contract called for delivery at an agreed-upon lower price. The power seller stopped delivering the promised electricity to the buyer, demanding more money for it. Naturally, the buyer objected, pointing out that the contract did not provide for changing the price just because market conditions changed. That was a risk that the buyer and seller agreed to take when they signed their contract. Still, the seller refused to deliver electricity. The seller was sued and defended itself vigorously so that legal costs racked up on both sides. All the while that bitter accusations flew back and forth, the seller kept offering to make a new deal with the plaintiff. The deal involved renegotiating their contract to make adjustments for extreme changes in market prices. The plaintiff resisted, always pointing—rightly—to the contract. But the plaintiff also really needed the electricity and couldn't get it anywhere else for a better price than the seller, my client, was willing to take—and my client knew that. Eventually, the cost of not providing the necessary electricity to their own clients became so great that the plaintiff caved in and took the deal they were offered.

Here was nasty, avaricious human nature hard at work in just the way game theorists think about it. Yes, there was a contract, and its terms were clear enough, but the cost of fighting to enforce the contract became too great. However much the plaintiff declared its intent to fight the case in court, the defendant knew it was bluffing. The plaintiff's need for electricity and the cost of battling the case out in court were greater than the cost of accepting a new deal. And so it was clear that the terms of the contract were not renegotiation-proof. The original deal was set aside and a new one was struck. The original deal really was not a firm commitment to sell (or probably, for that matter, to buy) electricity at a specified price over a specified time period when the market price moved markedly from the price stipulated in the agreement. Justice gave way, as it so often does in our judicial system, to the relative ability of plaintiffs and defendants to endure pain.

Commitment problems come in other varieties. The classic game theory illustration of a commitment problem is seen in the game called the prisoner's dilemma, which is played out on almost every cop show on TV every night of the week. The story is that two criminals (I'll call them Chris and Pat) are arrested. Each is held in a separate cell, with no communication between them. The police and the DA do not have enough evidence to convict them of the serious crime they allegedly committed. But they do have enough evidence to convict them of a lesser offense. If

Chris and Pat cooperate with each other by remaining silent, they'll be charged and convicted of the lesser crime. If they both confess, they'll each receive a stiff sentence. However, if one confesses and the other does not, then the one who confesses—ratting out the other—will get off with time served, and the other will be put away for life without a chance for parole.

It is possible, maybe even likely, that Chris and Pat, our two crooks, made a deal beforehand, promising to remain silent if they are caught. The problem is that their promise to each other is not credible because it's always in their interest—if the game is not going to be repeated an indefinite number of times—to renege, talking a blue streak to make a deal with the prosecutor. Here's how it works:

THE PRISONER'S DILEMMA

Pat's Choices→ Chris's Choices ↓	Don't confess (stay faithful to Chris)	Confess (rat out Chris)
Don't confess (stay faithful to Pat)	Chris and Pat get 5 years	Chris gets life; Pat gets time served
Confess (rat out Pat)	Chris gets time served; Pat gets life	Chris and Pat get 15 years

After Chris and Pat are arrested, neither knows whether the other will confess or really will stay silent as promised. What Chris knows is that if Pat is true to his word and doesn't talk, Chris can get off with time served by betraying Pat. If instead Chris stays faithful to her promise and keeps silent too, she can expect to get five years. Remember, game theory reasoning takes a dim view of human nature. Each of the crooks looks out for numero uno. Chris cares about Chris; Pat looks out only for Pat. So if Pat is a good, loyal buddy—that is, a sucker—Chris can take advantage of the chance she's been given to enter a plea. Chris would walk and Pat would go to prison for life.

Of course, Pat works out this logic too, so maybe instead of staying silent, Pat decides to talk. Even then, Chris is better off confessing than she would be by keeping her mouth shut. If Pat confesses and Chris stays silent, Pat gets off easy—that's neither here nor there as far as Chris is concerned—and Chris goes away for a long time, which is everything to her. If Chris talks too, her sentence is lighter than if she stayed silent while Pat confessed. Sure, Chris (and Pat) gets fifteen years, but Chris is young, and fifteen years, with a chance for parole, certainly beats life in prison with no chance for parole. In fact, whatever Chris thinks Pat will do, Chris's best bet is to confess.

This produces the dilemma. If both crooks kept quiet they would each get a fairly light sentence and be better off than if both confessed (five years each versus fifteen). The problem is that neither one benefits from taking a chance, knowing that it's always in the other guy's interest to talk. As a consequence, Chris's and Pat's promises to each other notwithstanding, they can't really commit to remaining silent when the police interrogate them separately.

IT'S ALL ABOUT THE DOG THAT DIDN'T BARK

The prisoner's dilemma illustrates an application of John Nash's greatest contribution to game theory. He developed a way to solve games. All subsequent, widely used solutions to games are offshoots of what he did. Nash defined a game's equilibrium as the planned choice of actions—the strategy—of each player, requiring that the plan of action is designed so that no player has any incentive to take an action not included in the strategy. For instance, people won't cooperate or coordinate with each other unless it is in their individual interest. No one in the game-theory world willingly takes a personal hit just to help someone else out. That means we all need to think about what others would do if we changed our plan of action. We need to sort out the "what ifs" that confront us.

Historians spend most of their time thinking about what happened in the world. They want to explain events by looking at the chain of things that they can observe in the historical record. Game theorists think about what *did not* happen and see the anticipated consequences of what didn't happen as an important part of the cause of what did happen. The central characteristic of any game's solution is that each and every player expects to be worse off by choosing differently from the way they did. They've pondered the counterfactual—what would my world look like if I did this or I did that?—and did whatever they believed would lead to the best result for them personally.

Remember the very beginning of this book, when we pondered why Leopold was such a good king in Belgium and such a monster in the Congo? This is part of the answer. The real Leopold would have loved to do whatever he wanted in Belgium, but he couldn't. It was not in his interest to act like an absolute monarch when he wasn't one. Doing some counterfactual reasoning, he surely could see that if he tried to act like an absolute ruler in Belgium, the people probably would put someone else on the throne or get rid of the monarchy altogether, and that would be worse for him than being a constitutional monarch. Seeing that prospect, he did good works at home, kept his job, and freed himself to pursue his deepest interests elsewhere. Not facing such limitations in the Congo, there he did whatever he wanted.

This counterfactual thinking becomes especially clear if we look at a problem or game as a sequence of moves. In the prisoner's dilemma table I showed what happens when the two players choose without knowing what the other will do. Another way to see how games are played is to draw a tree that shows the order in which players make their moves. Who gets to move first matters a lot in many situations, but it does not matter in the prisoner's dilemma because each player's best choice of action is the same—confess—whatever the other crook does. Let's have a look at a prospective corporate acquisition I worked on (with the details masked to maintain confidentiality). In this game, anticipating what the other player will do is crucial to getting a good outcome.

The buyer, a Paris-based bank, wanted to acquire a German bank. The buyer was prepared to pay a big premium for the German firm but was insistent on moving all of the German executives to the corporate headquarters in Paris. As we analyzed the prospect of the acquisition, it became apparent that the price paid was not the decisive element for the Heidelberg-based bank. Sure, everyone wanted the best price they could get, but the Germans loved living in Heidelberg and were not willing to move to Paris just for money. Paris was not for them. Had the French bankers pushed ahead with the offer they had in mind, the deal would have been rejected, as can be seen in the game tree below. But because their attention was drawn to the importance the Germans attached to

FIG. 3.1. Pay Less to Buy a Bank

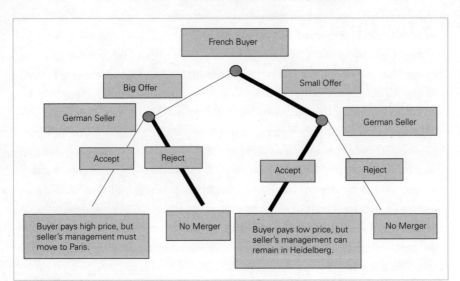

where they lived, the offer was changed from big money to a more modest amount—fine enough for the French—but with assurances that the German executives could remain in Heidelberg for at least five years, which wasn't ideal for the French, but necessary for their ends to be realized.

The very thick, dark lines in the figure show what the plans of action were for the French buyer and the German seller. There is a plan of action for every contingency in this game. One aspect of the plan of action on the part of the executives in Heidelberg was to say *nein* to a big-money offer that required them to move to Paris. This never happened, exactly because the French bankers asked the right "what if" question. They asked, What happens if we make a big offer that is tied to a move to Paris, and what happens if we make a more modest money offer that allows the German bank's management to stay in Heidelberg? Big money in Paris, as we see with the thick, dark lines, gets *nein* and less money in Heidelberg encourages the seller to say *jawohl*. Rather than not make the deal at all, the French chose the second-best outcome from their point of view. They made the deal that allowed the German management to stay put for five years. The French wisely put themselves in their German counterparts' shoes and acted accordingly.

By thinking about the strategic interplay between themselves and the

German executives, the French figured out how to make a deal they wanted. They concentrated on the all-important question, "What will the Germans do if we insist they move to Paris?" No one actually moved to Paris. Historians don't usually ask questions about things that did not happen, so they would probably overlook the consequences of an offer that insisted the German management relocate to France. They might even wonder why the Germans sold so cheaply. In the end, the Germans stayed in Heidelberg.

Why should we care about their moving to Paris when in fact they didn't? The reason they stayed in Heidelberg while agreeing to the merger is precisely because of what would have happened had the French insisted on moving them to France: no deal would have been struck, and so there would have been no acquisition for anyone to study.

The two games I have illustrated in the preceding pages are very simple. They involve only two players, and each game has only one possible rational pair of strategies leading to an equilibrium result. Even a simple two-player game, however, can involve more than one set of sensible plans of action that lead to different possible ends of the game. We'll solve an example of such a game in the last chapter. Of course, with more players and more choices of actions, many complicated games involve the possibility of many different strategies and many different outcomes. Part of my task as a consultant is to work out how to get players to select strategies that are more beneficial for my client than some other way of playing the game. That's where trying to shape information, beliefs, and even the game itself become crucial, and in this next section I'd like to show you just what I mean.

WANT TO BE A CEO?

As we all know, great jobs are getting harder to come by, and reaching the top is as competitive as ever. Merit may be necessary, but, as many of us can attest, it's unlikely to be sufficient. There are, after all, many more well-qualified people than there are high-level jobs to fill.

That being said, even if you've managed to mask or overcome your personal limitations and have been blessed with great timing and good luck such that you now find yourself in the rarefied air of the boardroom, there's something worth knowing that might have escaped you, something that might still prevent you from grabbing that cherished top spot: the selection process.

That's right, understanding and shaping the process by which a CEO or other leaders are chosen can tip the competition in your favor. It's funny that few of us pay much attention, in a strategic sense, to something as prosaic as how votes are counted, whether in the boardroom or national elections. And yet the method used to translate what people want into what they get can turn a losing candidacy into a winning one.[3]

When I talk about shaping outcomes based on voting, I don't mean anything like miscounting or cheating. I don't mean relying on hanging chads or anything like that. I'm just thinking about the many regular, commonly used ways of arriving at a choice based on what voters or shareholders or board members want.

Few board members or shareholders pause to think about how the votes are going to be counted when they select a new CEO. Hardly anyone asks whether it really matters if we require a candidate to get a majority or a plurality; if we count just votes for people's first choice or we allow them to express their first and second (or even more) preferences; if in decisions with many candidates we vote on all of them at once or we pair them up in head-to-head contests. And yet you can bet your bottom dollar that these decisions really can change the results.

Just think back to the hotly contested 2008 Democratic Party primaries. The Democrats allocated delegates from each state roughly proportionally to the candidates based on their share of the popular vote in each primary. Barack Obama won a majority of delegates that way, and was ultimately elected president. If the Democrats had used the Republicans' winner-take-all rule in each primary, Hillary Clinton would have won enough delegates to be the nominee, and she too probably would have gone on to beat John McCain. That's a pretty big consequence of a seemingly inconsequential rule.

There is, of course, no right way to count votes. Every method has advantages and disadvantages. So we might as well use voting rules when we can to help the candidates we favor. Generally we don't have the opportunity to change how votes are counted in government elections, but we sure do when it comes to corporate decisions.

In fact, I've twice used the range of boardroom voting procedures to help shape corporate choices of CEOs. Once the effort was entirely successful, and the second time, well, the candidate that my partner and I helped rose from obscurity to be treated as a very serious candidate. He ultimately lost, but he did so much better than anyone expected that he was quickly hired away from his company to become the CEO of a different firm—and he was a great success there.

How was the CEO selection process modeled? Let's take a look at my first experience on this front (it has the nice feature that even the person who was chosen as CEO didn't know—and probably still does not know—how he won). Here's what happened:

The retiring CEO of the company in question—certainly it must be obvious that there is a need for anonymity here—didn't have strong feelings about who he wanted to replace him. He did, however, have very strong feelings about who he did *not* want to replace him, and it just so happens that the person he didn't want was the leading candidate for succession. The retiring CEO truly despised this person, who had been his nemesis for many years, and he hired me in secret to help engineer the CEO selection. The modeling job: figure out how to beat the detested front-runner.

As with any analysis, the first step was to figure out just what were the real issues that had to be resolved. In this case, the big questions were simple enough. They required working out who the prospective candidates were and how they stacked up against each other. Let's call the candidates Larry, Moe, Curly, Mutt, and Jeff—with Mutt being the guy to beat.

The problem was best analyzed with a bunch of beauty contests. Each beauty contest asked how the stakeholders with a say in selecting the CEO felt about Larry vs. Moe, Larry vs. Curly, Larry vs. Mutt, Larry vs. Jeff, Moe vs. Curly, and so forth.

Once the issues—the beauty contests—are specified, we need to know each stakeholder's position, or which candidate he or she favors in each head-to-head contest, and by how much. (I will go into further detail in subsequent chapters as to the particular methodology behind these assessments.)

Thankfully, in this particular case, we have a good source of information: the outgoing CEO. He knew who the players were and he knew how they felt about each candidate. And you can be assured he didn't get to be CEO without knowing which of his colleagues had real clout and who would just go along with the wishes of others.

The existing procedure for succession in the CEO's company did not involve head-to-head contests, ranking of candidates, runoffs, or a host of other common voting rules. Instead, they normally voted for all the candidates simultaneously, just as is done in American presidential elections. Whoever got the most votes would be the winner. Now that would have been very bad news for my client, the retiring CEO. It was clear that such a procedure, a fine, upstanding, legitimate procedure, would result in the election of Mutt, the detested candidate. What to do?

The first thing was to sort out who was likely to win each of the beauty contests. The stakeholders consisted of the members of the company's CEO selection committee. Let's say the committee was made up of fifteen individuals, with one vote each. The outgoing CEO's information on the comparisons of candidates in pairs allowed me to tease out the strict order in which different committee members preferred the candidates. It was apparent that the fifteen committee members divided equally into five voting blocs based on their preferences, with three members in each group. Here are the five different preference orderings held by the members of the selection committee, with candidates listed from most preferred to least preferred:

1. Mutt, Jeff, Larry, Curly, Moe
2. Mutt, Moe, Curly, Larry, Jeff
3. Moe, Mutt, Curly, Larry, Jeff
4. Jeff, Moe, Curly, Larry, Mutt
5. Larry, Jeff, Curly, Moe, Mutt

In a contest in which everyone got to vote just one time, such as is used in the United States to pick the president, the detested Mutt would get 6 votes (two blocs held him in first place and each had three members), Moe 3, Jeff 3, Larry 3, and poor Curly 0. Mutt wins. That is exactly what had to be prevented.

However, under another voting system, if committee members got to cast 4 votes for their first choice candidate, 3 for second, 2 for third, 1 for fourth, and no points for their last choice (a method known as a Borda count), then Mutt and Moe would receive 33 weighted votes each, Jeff would get 30, and Larry and Curly would bring up the rear with 27 each. If they then held a tie-breaking runoff between Mutt and Moe, Moe would pick up the votes of the third, fourth, and fifth blocs. They each favored him over Mutt. Moe would be the new CEO by this procedure.

So already we can see that there is a rule that could beat Mutt. However, it was clear enough that this procedure would be tough to get through the committee. It was just too complicated to ask members first to rank candidates and then to hold a runoff once they discovered there was a tie. Such a convoluted election process would easily arouse suspicion among committee members. They might have wondered why the retiring CEO was asking them to do something so elaborate when they could just vote straight up for any candidate they wanted.

Even if this complicated procedure could get committee approval, we would not have been home free. The procedure itself might be thwarted if a Mutt supporter caught on. For instance, if even one member of the second bloc worked out the results ahead of time, that member—who really wanted Mutt—could strategically (that is, by lying) decide to rank Jeff second and Moe last. Sure, that would have been a misrepresentation, but then the voter's interest was in the final choice, not any intermediate decision. Acting strategically by inflating Jeff's ranking, the bloc member would have ensured that Moe's total in the rankings would come to 30 instead of 33, and so would Jeff's, greatly complicating the process by creating a tie for second place and increasing the odds that Mutt would win. After all, in a runoff against Jeff (likely since more blocs preferred Jeff to Moe than the other way around), Mutt would be the winner. By acting strategically, then, it was possible for one or more members of the second bloc to ensure the election of their most preferred candidate, the detested Mutt. That was a chance I wasn't willing to take. Instead I decided to get Curly elected.

Poor Curly—he was at a huge disadvantage. No one viewed him as their first-place choice. Nobody even thought of him as second choice. In fact, he was barely on anyone's radar screen. I know that for sure because after working out how to get him elected, I had a conversation with a member of the selection committee. My role in the process was a secret, known only by the then CEO and me. I asked the committee member who he thought would be chosen, and he mentioned Mutt and maybe Moe. I nonchalantly asked about Jeff, Larry . . . and Curly. He took Jeff and Larry seriously, although he didn't think they could win. Then he told me that neither he nor anyone else on the committee understood why

Curly had put himself forward. After all, he said, he just doesn't have a prayer, no one favors him. Sure, they liked him well enough, but they just didn't seem to think of him as CEO material. Curly's relative obscurity was, in this case, his great advantage. It was unlikely that anyone paid enough attention to Curly's candidacy to maneuver strategically to thwart his prospects, since they didn't think he had any.

Okay, so now the fun begins. The outgoing CEO was well liked and highly respected. He had done a good job. The beauty contests revealed enough to show how to get Curly elected (do you see how?), but I needed to analyze one more issue first. The question was whether the retiring CEO had enough clout to persuade the selection committee to follow the winning voting procedure. The analysis of that question showed that indeed he could get the committee to follow the voting rule he suggested, provided it wasn't too complicated. Fortunately, the procedure my analysis suggested was an eminently reasonable rule. It wasn't particularly complicated, and it capitalized on the committee's majority not being keen to elect Mutt in the first place (remember, he had 6 first-place votes; 9 first-place votes were distributed among the others).

Agenda control—determining the order of decision making—can be everything. In this case it was. By setting the right agendas we could create a series of winning coalitions, each made up of different members from the one before, ending with a winning coalition supporting Curly and leaving no other candidate up for consideration.

The committee members understood that the real contest was between Mutt and Moe—or so they thought. To reinforce their view, the outgoing CEO persuaded the committee to use an agenda—a sequence of choices—that was made up of a specific sequence of head-to-head elimination contests. Of course there were too many candidates to ask the committee to compare each candidate to each other candidate, two at a time. That would have meant ten votes. Instead, the retiring CEO persuaded the committee to vote on Mutt versus Moe, with the loser of that contest being eliminated from consideration and the winner then going up against Jeff. Whoever lost that contest would be dropped from consideration, and the winner (who at this point in principle could have been Mutt or Moe or Jeff) would then be voted on against Larry, and the winner of that vote would finally be voted on against Curly. Whoever was left standing after those four votes would be deemed the winner.

This seemed like a good idea to the selection committee. They thought that by leading with strength—Mutt vs. Moe—they would quickly arrive at the one of those two who overall was most desired as CEO. How wrong they were. To be sure, anyone paying close attention to the five voting blocs' preferences could have worked out how the retiring CEO's agenda would work out, but it was unlikely that the committee members knew the full preference ordering of their compatriots. They, after all, were unlikely to conduct the sort of expert interviews called for by the model. Since they weren't asked to announce their candidate rankings, the rule proposed by the retiring CEO did not compel them to reveal to each other their full ranking of candidates. Probably on their own they had not probed one another beyond second-place preferences. That, presumably, was why they paid so little attention to Curly. So here is what happened:

Moe beat Mutt right off the bat—by a vote of 9 to 6 (blocs 1 and 2 voting for Mutt and the rest voting for Moe, as you can see from the rankings listed earlier). Mutt, being the loser, was dropped from consideration under the seemingly reasonable supposition that more people wanted Moe than Mutt (9, as we saw, to 6). Fair enough. Everything after that was gravy, because my client's main concern was to beat Mutt. But then my client also liked the idea of choosing Curly. He thought that would make him look even better in retrospect, and besides, he was fond of Curly and thought being CEO would be a nice way to cap Curly's career.

The selection committee then considered Moe and Jeff in accordance with the agreed-upon agenda. Jeff beat Moe as handily as Moe had beaten Mutt. Bloc 1 wanted Mutt most of all, but now, confronted with a choice between Moe and Jeff, they went for Jeff. He was their second-place choice, while Moe came in last for bloc 1. Blocs 4 and 5 also thought Jeff was a better prospective CEO than Moe. Only blocs 2 and 3 favored Moe over Jeff. That gave Jeff 9 votes to Moe's 6. Mutt having already been eliminated, no one on the committee stepped back to ask what would happen if they took the opportunity to choose between Jeff and Mutt. As you can see from the bloc preferences, there was yet another winning coalition (blocs 1, 2, and 3) with whose support Mutt would have beaten Jeff, but again, Mutt had been taken out of the picture by Moe in accordance with the elimination rules agreed to.

Mutt and Moe, the apparent front-runners, were now out of the race. Moe beat Mutt, and Jeff beat Moe. Jeff, Larry, and Curly were still standing. Jeff and Larry each had first-place supporters, so they were run against each other next. Blocs 2, 3, and 5 favored Larry over Jeff. Jeff was out, leaving a final choice between Larry and Curly. Of course you can easily see that Curly is going to defeat Larry. Blocs 2, 3, and 4, in the ever-shifting winning coalition of voters, favored Curly over Larry. Curly, being the last man standing, was the new CEO much to (almost) everyone's surprise. Still, they felt the process had been fair and square, and in its own way it was.

No one seemed to notice that the agenda had decided the outcome. It so happens that Larry was the only candidate that Curly could have beaten. If the agenda had been different, Curly would have lost. Just as Curly could not beat anyone other than Larry, so too could Larry not defeat anyone other than Jeff. Moving Larry up in the agenda would have wiped him out and Curly with him. In fact, because preferences went around in circles (or, to put it technically, they were intransitive), an agenda could be put forward to make any of the candidates into the winner fair and square.

When the vote was over, the committee member with whom I had talked earlier invited me to lunch. He had one question for me: Did you have anything to do with picking our CEO? I smiled and changed the subject. He was sure I did, and I knew he knew, but I was sworn to secrecy. The lunch was great.

SORRY, EINSTEIN: GOD DOES ROLL THE DICE

As we see in moving from the prisoner's dilemma to the bank example to the voting strategies, even in relatively simple games involving relatively few players there can be multiple outcomes. This fact adds yet another strategic dimension to decision making, particularly as games, in the real world, are often played over and over between the same players.

Any time a game has more than one possible result, there is a special type of strategy (called a mixed strategy) that can influence what happens. In a mixed strategy, each player chooses actions probabilistically—say, by rolling dice—to influence what other players expect to get out of the game. Einstein's God may not have played dice with the universe, but we mortals definitely roll the dice with each other.

Whenever you watch a football game and complain about a coach's choice of plays, you probably were watching a mixed strategy at work. For instance, when the ball is on the one-yard line, the play that is most likely to get the ball across the goal line is for the fullback to jump over the pile of players in front of him. Yet coaches often have the quarterback pass the ball or hand off to a running back. The reason: if a coach always called for the fullback to go over the top, the defenders would concentrate the defense at that point, and the play would probably fail. By mixing the calls, the offense forces the defense to spread out, thereby improving the odds of success over repetitions of the situation. Interestingly, this sort of mixing of strategies carries important lessons for business, politics, and lots of other parts of life. Rolling the dice is one way to alter how other people perceive a situation.

Using strategies that involve mixing up moves to create a change in expectations is something that comes up all the time. Although applied game theorists often like to ignore these complicated "mixed strategy" approaches to problems, they do so at their own peril. Rolling the dice can really make a difference in how things turn out.

Examples of such gambling are all around us, and some great movies roll the dice very cleverly to create climactic moments. Who can forget in *The Princess Bride* the back-and-forth over which wineglass is poisoned, and the clever resolution (both were poisoned—it pays to build up an immunity to the poison you plan to use on yourself and others). Or how about the fabulous scene from *The Maltese Falcon* in which Sydney Greenstreet's character, Kasper Gutman, desperately wants the jewel-encrusted bird? Only Sam Spade (Humphrey Bogart) knows where it is, and Sam Spade is no fool. Gutman threatens that Joel Cairo (Peter Lorre) will torture Spade to find out where the bird is, but Spade counters, "If you kill me, how are you going to get the bird? If I know you can't afford to kill me till you have it, how are you going to scare me into giving it to you?" Here Spade, like any good game theorist, questions the credibility of Gutman's commitment to make him talk. We know and Sam Spade knows that, without a real commitment to kill him, Gutman can't get him to talk. But Gutman is no fool either. He knows just how to make the dice tumble, creating the prospect that Spade will talk to save himself. After some clever give-and-take, Gutman retorts, "As you know, sir, men are likely to forget in the heat of action where their best interests lie and let their emotions carry them away."

There it is: "Men are likely to forget in the heat of action where their best interests lie and let their emotions carry them away." How beautifully put. He's just explained that Joel Cairo will try to be careful not to kill Spade, but then, Cairo can get emotional, so there is also a chance that if Spade doesn't talk he'll end up dead. In this brief exchange we see three lovely principles of game theory at work: the question of credible commitment; the use of playing probabilistically to alter how others look at the situation; and the pretense of irrationality (the heat of the moment) for strategic advantage. What could be truer to life's fears and calculations? How many of us would dare to stay silent given Sam Spade's gamble: keep the bird and maybe die, or give up the falcon and (maybe) live?

With a bit of luck, it will become apparent that game theory is not limited to parlor tricks, movie scripts, and brainteasers. It is a powerful tool for reshaping the world. In the remaining chapters, we will use these foundations to see just what kind of problems rational choice theory can tackle, and how math, science, and technology now allow us to predict and engineer particular outcomes that we might otherwise assume would only be determined by a random mix of good or bad fortune and a heavy dose of human whim.

4

BOMBS AWAY

WE NOW UNDERSTAND the basic principles of game theory. That's all well and good, but how do we use these principles to solve the big problems of our time? Well, let's not mess around—let's take a look at North Korean disarmament.

Being born in North Korea pretty much ensures having a miserable life. The average North Korean has to work an entire year to earn what an average American, Irish, or Norwegian worker makes in around four days. Money isn't everything, but it is a pretty good first-cut approximation of quality of life.

Of course, being born in North Korea is not always bad news for everyone. Kim Jong Il, that country's "Dear Leader," seems to do very well indeed. He is estimated to be worth around $4 billion. That's about one-third of North Korea's annual gross domestic product. (Poor Bill Gates, his wealth is only about 0.4 percent of the U.S. gross domestic product.)

Kim's wealth, womanizing, heavy drinking, and gourmand tastes incline many to think of him as an inconsequential, even frivolous, lightweight dictator. He is frequently described as irrational, erratic, and dangerous. Most assuredly he is the last of those, but irrational and lightweight—I think not. Sure, he came by his position the old-fashioned way—he inherited it (on the death of his father, the "Great Leader," Kim Il Sung). Still, that was a long time ago. No fool stays in power for years on end when there are so many generals, sons, and wives waiting in the wings to launch a coup.

Kim Jong Il is a savvy, skillful, vicious demagogue. If he is erratic it is because it serves his interests. While we might be inclined to laugh at this seemingly odd little New Age dictator, he cleverly maneuvers within the limitations of the miserable cards he's been dealt to make himself a menace on the world stage. All the while that he engages in fomenting terror at home and abroad, he continues to rule a country that he and his father made dirt poor.

When Kim came to power in 1994, North Korea had virtually nothing to sell in the world. Its stature was about as low as it could get. So miserable has he made life in North Korea that many of its hapless citizens have been reduced to eating tree bark. Maybe as many as 10 percent of these poor souls, while taught to revere Kim Jong Il as a god, have died from starvation in the last decade. Yet today, North Korea is on everyone's radar screen. Why? Because while there is hardly an economy to speak of, there is the business of missile and weapons development that Kim has carved out to remarkable success. His people may starve, but he has put the money he takes from them to work, turning North Korea into a nuclear menace. His potential to launch nuclear-tipped missiles surely provides food for thought for the world even if he puts no food on the average North Korean's dinner table. Sure, Kim's government is reviled in almost every corner of the globe, but just about everyone takes notice. They didn't a couple of decades ago.

Today, the governments of the United States, China, Russia, Japan, and the Republic of Korea try to work out how to bring Kim Jong Il's rogue state into the mainstream of international affairs. For years he was threatened, he was coaxed, he was urged to be better, but he also was rewarded by Bill Clinton, at Jimmy Carter's behest, in the hope that North Korea's behavior would improve. The rewards Kim's regime got—in the form of foreign aid, technology transfers, and food relief—came with little or no genuine commitment on Kim's part, just empty promises. More recently, the six-nation talks seem to have made progress. The task of diminishing his threat appears to be moving forward, although in fits and starts. Finding a path to resolving his nuclear threat can be, and perhaps has been, advanced by accurate predictions and engineering aimed at getting Kim to put his bombs away.

In early 2004 the Department of Defense hired me as a consultant to investigate alternative scenarios for trying to get North Korea to behave better on the nuclear front. I can only sketch the solutions I came up with here, but even so, we can see how to think about such issues. I will focus attention on the scenario that held out the greatest promise for solving the problem. This scenario looks at trade-offs between U.S. political and economic concessions in exchange for North Korean concessions on the nuclear front. Before delving into details, however, let me be perfectly clear that I am taking and seek no credit for whatever progress may have been made. Policy consultants almost never know whether they are being listened to. They almost never know what other sources of advice are being heard or what those sources are saying. All I can do is report what my analysis showed and relate it to what eventually happened.

What kind of information is needed to make reliable predictions and tactical or even strategic recommendations? First, of course, it is essential to define the questions for which we want answers. A question like "How can we get Kim Jong Il to behave better?" is too vague. We need to define the objective more precisely, and we need to know the range of choices that Kim and his government can undertake. In this case, the choices included the possibility that Kim and his regime would refuse to negotiate about nuclear weapons at all; would negotiate but then cheat on any agreement as soon as doing so became advantageous (Kim's preferred approach); would slowly reduce the extent of North Korea's nuclear program in exchange for different levels of U.S. economic and security concessions; would eliminate the program conditionally (with various gradations of conditions having been specified); or would eliminate the nuclear program unconditionally (the most preferred outcome from the perspective of the American president and his foreign policy team).

Next we want to know what background conditions should be imposed on the question. For example, we might ask which of the above policies Kim Jong Il could be induced to adopt if the United States publicly targeted nuclear-tipped missiles at North Korea, or if the United States guaranteed North Korea's security within its borders, or a host of other possibilities. Each condition defines a scenario so that we can compare what is likely to happen if the United States (or some other government) takes this or that action. This way, we start to answer "what if" questions. Once the issue, the options, and the scenarios are defined, then only a very few facts and a bit of logic are needed to identify solutions.

First, the facts. In my experience, all that is necessary to make a reliable prediction is to:

1. Identify every individual or group with a meaningful interest in trying to influence the outcome. Don't just pay attention to the final
   decision makers.
2. Estimate as accurately as possible with available information what
   policy each of the players identified in point 1 is advocating when
   they talk in private to each other—that is, what do they say they
   want.
3. Approximate how big an issue this is for each of the players—that
   is, how salient is it to them. Are they so concerned that they would
   drop whatever they're doing to address this problem when it comes
   up, or are they likely to want to postpone discussions while they
   deal with more pressing matters?
4. Relative to all of the other players, how influential can each player
   be in persuading others to change their position on the issue?

That's all you need to know. *That's all?* you might ask. What about history? What about culture? What about personality traits? What about
almost everything else that most people think is important to know?
Knowing all of those things would be great, and I will say more about them
in a moment, but none of that information is crucial to making correct
forecasts or to engineering policy change. Sure, it helps. It is generally better to know more than less. Still, anyone who does not put together information on the four factors I listed above is unlikely to assess a situation
correctly.

What's interesting about the four pieces of information that I contend
are crucial is that while they're not the kind of stuff that can be easily
looked up in a book, it *is* possible to tease such information out of articles
in *The Economist, US News & World Report, Time, Newsweek, The Financial Times, The New York Times, The Wall Street Journal,* and from Internet
stories and other news outlets. Understanding the availability of such information, and having the confidence to employ it, is a big part of predicting and engineering outcomes. Admittedly, it is a lot of work going through
so many news sources, and for problems with a short fuse, that approach
can take too long. Luckily, there is a more efficient way to get the information—ask the experts. It's that simple.

Experts have invested years in learning a place's culture, language, and
history. They follow the intimate political details that go on in the area

they study. If anyone knows who will try to shape decisions, how influential those people can be, where they stand, and how much they care about an issue, it is the experts. Come to think of it, isn't knowing this information what it means to be an expert?

About now you might wonder, if the experts know the information needed to make predictions, what do we need a predictioneer for? Here is where specialization of skills is really important. It's important to remember that experts alone do not do nearly as well at anticipating developments as do experts combined with a good model of how people think. A declassified CIA study reports that my forecasting model has hit the bull's-eye about twice as often as the government's experts who provided me with data.[1] I certainly don't know more than they do about the countries or problems they study. In fact, I often know no more than what they tell me. But they're not experts on how people make choices, because that, after all, is not the focus of their knowledge.

With all of the information we collect in order to predict and shape outcomes, computer modeling is necessary both to organize the data and to run simulations of negotiations or exchanges. Think of these simulations as a game of chess in many dimensions, in which the computer calculates everyone's expected actions, taking anticipated responses by everyone else into account. The computer has a tremendous advantage over experts, analysts, or the smartest decision makers when it comes to playing such a complicated game. Computers don't get tired; they don't get bored; they don't need coffee breaks or much sleep; and they have fabulous memories. They are content to crunch as much information as we shovel into them.

Consider the computer's advantage. Suppose we were examining the North Korean nuclear problem in 2004, as I was, and suppose we simplified it (which I didn't) to consider just five players: George W. Bush, Kim Jong Il, Russia's Vladimir Putin, China's Hu Jintao, and South Korea's Roh Moo Hyun (ignoring Japan for the moment). How many conversations among the parties to the talks might each of those five decision makers want to know about?

George Bush certainly would want to keep track of what he said to each of the other four and what each of them said to him. They would all want to keep tabs as well on any proposals they made and any they heard from others. That's twenty exchanges of views right there. Certainly that is not all that any of them would want to know. Bush would be interested to know, or at least try to figure out, what Kim Jong Il might be saying to Putin, to Hu Jintao, and to Roh, and each of them would want to know about the conversations they were not directly involved in too. That's another sixty possible discussions. And Bush might even want to know what, for example, Putin thought Kim was saying to Hu Jintao and to Roh Moo Hyun, not to mention what he thought Roh said to Hu and to Kim, and so on.

All in all, taking all the layers of information being traded back and forth among just these five decision makers, there are 120 possible exchanges or imagined exchanges (that is, 5 factorial, or $5 \times 4 \times 3 \times 2 = 120$) to know about. Keeping track of those 120 possible offers and counteroffers that might be on the table is essential in sorting out what is best to do at any moment in a negotiation. Those 120 possible exchanges of points of view and beliefs about such exchanges are what can happen in a single round of bargaining with just five stakeholders. It might be surprising to know that a smart person can keep that amount of information pretty straight in his or her head. Keeping the information straight, however, becomes an acute problem as the number of interested parties rises.

Just adding Japan's prime minister—the talks are, after all, six-party talks, not five-party talks—to the mix inflates the number of important bits of information six times from 120 to 720. Moving up just to ten players, the number of useful pieces of information rises astonishingly to over 3.6 million! No one—not Newton, not Einstein, not von Neumann—can keep that much information straight in his head; but of course the tireless computer can.

Alas, the computer's great memory and excellent work habits come at a price. There is a vital gap between how experts or newspaper articles express facts and how computers ingest and digest them. Like the rest of us, experts communicate in sentences. Models talk in numbers. So part of my job is to turn sentences into numbers so that the computer can crunch away. Numbers have big advantages over words—and not just for computers. Most importantly, numbers are clear; words are vague. It's essential to turn information into numerical values, and in fact it's not especially hard to do.

To get a sense of how readily experts know the information needed to make reliable predictions, and to see how easily the information can be turned into numbers, try an experiment. Interview someone you think of as really knowing about your friends and family, including perhaps yourself.

Pick an issue that is important to your family or friends. It doesn't have to be about world affairs; it could be about where to have dinner, or what movie to see, or whatever else leads to disagreements. The easiest sort of issue to do as a first try is what I call a beauty contest. Say you and some friends are trying to choose between two movies. Anyone who really truly wants to see *The Sound of Music* (or fill in whatever first-run movie might grab your fancy) gets a value of 100, and anyone really committed to seeing *A Clockwork Orange* (another great old film) gets a 0. Then you should be able to rate how strongly each friend leans toward one movie or the other. Any who are truly indifferent get a 50; anyone leaning slightly toward *A Clockwork Orange* might be close to 50, say at 40 or 45, and so on. You or the "expert" you are interviewing need to calibrate their strength of feeling as accurately as possible. This way, movie preferences are turned into one value for each chooser—that is, each family member and/or friend involved in shaping the decision. The process is exactly the same, although the choices may be more complex, whether deciding what movie to see or addressing North Korea's nuclear choices. Sure, the stakes differ, but once the essential facts are extracted, the process of turning stated objectives into a predicted outcome is the same.

Now estimate how eager each friend or family member—each player—is to weigh in on the decision. If you think a family member will drop what he or she is doing to discuss the movie to see, rate that person's "salience" (variable 3 on my list) close to 100 (no one is ever really at 100). The less focused you think someone is on the movie choice, the lower the salience score. If a family member is the sort who would say, "Look, I'll go to whatever movie you choose, but really I don't have time to get involved in picking which one," that's somewhere around a 10. On the other hand, if you think a friend will say, "I'm busy right now but call me back in ten minutes," that's pretty high salience. "Call me back in an hour" is lower, and "Call me back next week" is *much* lower. With a bit of effort, it shouldn't be that hard to calibrate how important the movie decision is to each person compared to the other decisions they have to make (not compared to each other, mind you, but compared to other things they need to do or deal with).

Finally, figure out who you think has the most influence among your friends or family members if you assume that everyone thinks the choice is equally important. Give the person credited as being most persuasive a score of 100 and rate everybody else relative to that. So if Harry is 100 and Jane is 60 and John is 40 in potential clout, and Jane and John want to see *The Sound of Music* and Harry wants to see *A Clockwork Orange*, then that means that John and Jane together just offset Harry's ability to persuade if they all care equally intensely about choosing a movie. If Jane were 60 and John were 70, then, all else being equal, they could persuade Harry to moderate his view and give more consideration to the movie Jane and John prefer. Of course, if someone else supported Harry's choice, that might create a strong enough coalition to defeat John and Jane. The dynamics get complicated, but the basic idea should be straightforward.

It's important to note that most of us make these assessments of interests in any situation. We just do such calculations naturally with relative judgments of where people stand on a given issue. What I've sketched above is simply a formalization of that natural process—which becomes all the more needed the more complicated the problems in question become.

By now you are probably thinking, Sure, people can fill in numbers to the questions, but it's just guesswork. Ask two experts the same question and you'll get two different answers. Guess what—that's not true. If it were, then there is barely any chance that a model like mine could achieve any consistent accuracy. There would be too much luck involved. In fact, the CIA has checked out the risk that different experts give greatly different answers leading to greatly different predictions. They found little variation in the predictive results from the sort of modeling I do, even when the people asked had dramatically different access to information. Academic experts, for instance, generally do not know the classified information that intelligence analysts have access to. Yet both groups tend to provide data so similar, wherever it's from, that the results hardly change when moving across these experts. Even more surprisingly, the answers often don't change much when the inputs for the computer model are put together by undergraduate students with no expertise at all.

Once I was teaching an undergraduate class at the University of Rochester while also investigating how best to get Ferdinand Marcos to resign as head of the Philippine government and create an atmosphere ripe for a free election in that country. William Casey, then Ronald Reagan's director of intelligence, asked me to study this problem, and I was locked in a (cold) lead-lined vault at CIA headquarters to do it. I had access to classified information, but was not even allowed to read my own report when it was finished. The report was for the eyes of only the president, the vice president, the secretaries of state and defense, the national security adviser, and a few others. Meanwhile, my students worked on the same problem and were given access to the computer program I had developed to help solve such problems. They extracted the required information from magazines such as *The Economist* and newspapers such as *The New York Times* and fed their data into the computer. Ninety percent of them arrived at the same conclusions I reached in the lead-lined vault, and those conclusions about strategy proved to work rather well. This should tell us that the information needed to make good predictions is not terribly exotic; it does not require years of learning some other country's language, history, and culture, although all of that is a big help. It should also tell us that a lot of classified information is easily reproduced from open, public sources for those who are willing to work at it.

Now that we have an idea of where and how to find our information, working out how to get what any given player wants is the key to generating predictions, and ultimately to engineering outcomes. Since everyone involved in a given problem is concerned with getting what they want, their behavior and choices are predictable. Each and every one of them will act so as to lead to the attainable outcome that is closest to what they want, given what they believe about the situation.

What might people's goals be in a generic sense? Whatever the specific issue, I always operate under the assumption that everyone wants two things when making a decision (although different people weight those two things differently). One thing they want is a decision that is as close as possible to the choice they advocate. The second thing they want is glory—the ego satisfaction that comes from the recognition by others that they played an important part in putting a deal together.

Some people care so much about getting credit for putting an agreement together that they're willing to shift their position dramatically if that will help promote a deal. Others prefer to go down in a blaze of glory, backing a losing position rather than making concessions that would make a deal feasible. Everyone shares these two goals: get their preferred outcome, and get credit for any outcome. Different people value one or the other differently, and so they're willing to trade away returns on one dimension to get better returns on the other.

Let's return to the North Koreans. After interviewing experts and compiling research, we know three important pieces of information about each player: what they say they want, how much they care, and how influential they can be. In the case of North Korea, the range of policy choices is depicted in figure 4.1, both in terms of substantive meaning and numerical value. (One easy way to get numbers out of policy stances is to ask experts to make marks on a line for several policy options, emphasizing that they should be spaced to reflect how close or distant the choices are substantively from each other. Then a ruler can be used to measure the distances, and voilà, a simple numeric scale has been created.)

My 2004 study of North Korea identified more than fifty players in this complicated international game. Two, Kim Jong Il and George W. Bush, had a veto, which meant that no deal could be struck without their support. Kim Jong Il's preferred position was to agree to a deal but to structure it so that he could cheat later, reneging on his promises (10 on the scale). According to the experts, Bush wanted the unconditional elimination of North Korea's nuclear program (100 on the scale). Without some strategizing, then, the two sides were unlikely to reach agreement, since the two most important decision makers were miles apart and both were thought to be reluctant compromisers. Yet a first approximation of the likely outcome suggested strong support for a consequential reduction in North Korea's nuclear capabilities accompanied by significant U.S. con-

FIG. 4.1. North Korean Issue

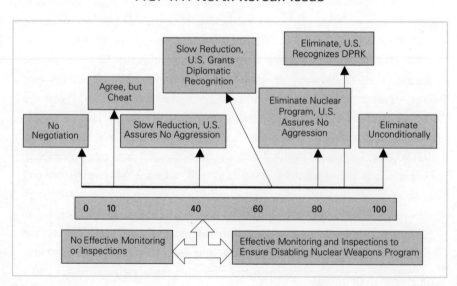

cessions, including security guarantees for North Korea and considerable foreign economic assistance. How did I arrive at that inference?

The information collected about each player in the North Korea game— their position on the issue, their salience for the issue, and the clout they could bring to bear—allows us to see how much power there is behind each possible outcome. Potential influence, one of the three pieces of information, tells us how persuasive each player could be, but not how persuasive each is. That depends on their willingness to apply their influence to the problem, and that in turn is determined by how salient the issue is. So we can define a player's power—the pressure they really exert to shape the outcome—as equal to their influence multiplied by their salience.

Using that information, we can formulate the first of two pretty reliable first cuts at a forecast (ignoring vetoes for now). Figure 4.2, on the next page, shows the distribution of power in support of each of the major policy stances on the issue scale. Think of it as showing how much power backs each option. It is like a map of mountainous terrain, with the positions garnering the most powerful support forming high, prominent peaks, and the positions with little support amounting to not much more than molehills. This picture of the power terrain is based on the answers to the questions posed to the experts about position, salience, and potential influence.

We can extract two insights from figure 4.2 that can help us predict if we are looking at a purely international political decision (which we are not). First, there really isn't that much clout supporting North Korea's nuclear program. None of the positions that favor North Korea's keeping a nuclear program are backed by a really large amount of power. Second, piling the mountains of power one on top of the other as we move across the spectrum of choices, we discover that the pile does not reach a majority of all the power until we get to the position designated as "Eliminate Nuclear Programs / U.S. Concessions."

Since we are continuing to assume, as we did back in Game Theory 101, that stakeholders prefer positions closer to their own to positions farther away, our first first-cut prediction is "Eliminate Nuclear Programs / U.S. Concessions." Why? Because if it takes a majority of power to enforce an outcome, then the winning position is the one for which the total of power to the left of it is less than 50 percent and the total of power to the right of it is also less than 50 percent. The only position for which that is true is "Eliminate Nuclear Program / U.S. Concessions."[2]

FIG. 4.2. The North Korean Nuclear Power Landscape

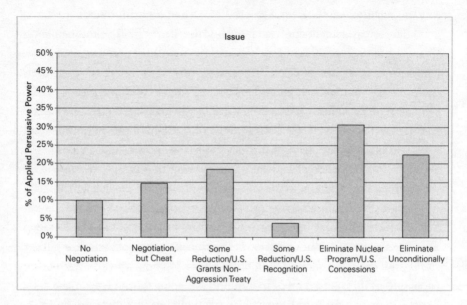

This prediction has some important limitations. For one thing, it ignores vetoes. Since it is not even close to Kim Jong Il's desired outcome (and is not all that close to the United States position either), we can be confident that North Korea will reject it unless there is sufficient political pressure on Kim Jong Il to get him to change his position. So we see we really must think through not only the international aspects of the issue but also the domestic dynamics in North Korea. To be sure, we need to work out whether it is possible for foreign stakeholders (like the U.S. president or the South Korean president) to change Kim's mind. But we also have to analyze the internal political pressure Kim might face if he resists or if he concedes the proposed solution by outsiders. These are some of the things my computer model does that are difficult to do in our heads.

We can look back at the information we collected from experts and assemble a second preliminary prediction. Again we will have to be mindful that we are ignoring the possibility of vetoes for the moment, as well as the dynamics that lead players to alter positions in response to credible threats and promises.

Rather than just look at the 50 percent break point in the power landscape, we can apply a different method to arrive at a sensible preliminary prediction. Let's multiply the influence of each player (calling influence I) by his or her salience (S) and multiply that result by the numerical value of the position each player advocates (P), then add those totals up for all of the players and divide that total by the sum of the influence times salience for each of the players (sum of I × S × P)/(sum of I × S). Now we have computed a value called a weighted mean. This is, roughly speaking, an average of what people, with their influence and commitment to a given issue taken into account, want. With the answer to this calculation in hand—it is 59.8—refer back to figure 4.1. There we see numerical values associated with possible solutions to the issue, and so we can see that 59.8 is equivalent to the position designated as "Slow Reduction, U.S. Grants Diplomatic Recognition." The details behind the calculation leading to this prediction can be found in the first appendix.[3]

Now we have two first-cut ways to predict what is likely to happen. Taking the two together, we can be fairly confident (still ignoring vetoes) that the solution to North Korea's nuclear problem lies somewhere between the initial, majority-power approximation (about 80 on the issue scale in figure 4.1) and the weighted mean position (60 on the issue scale). That easily, we have created pretty reliable initial forecasts for this issue—a narrow range within which a resolution is likely to be found if no one exercises a veto. The initial forecasts mean substantively that at the outset there was a possible resolution supporting a slow reduction in North Korea's nuclear capability followed by U.S. diplomatic recognition of the Democratic People's Republic of Korea. I will say more about that in a moment.

The really simple power-majority view coupled with the slightly more complicated weighted-mean view offers a good answer to the range of likely agreements, but they do not result in the best prediction possible. That takes a computer program to calculate the solution to a game in which players make policy proposals and try to exploit each other's egos to alter stances, climb over or reshape the mountains of power that lie in their way, and get the outcome they want. Still and all, there's now enough information to make a pretty reliable prediction. This baseline forecast is likely to be right around 70 to 75 percent of the time.[4]

Take a look back at the hills and valleys in figure 4.2, remembering that we are looking at the lay of the land in 2004. Doing so helps us understand that being really powerful does not assure success. The tallest mountain, the biggest mass of power, supports the idea that North Korea

should completely eliminate its nuclear program in exchange for concessions. That mass of power supports an outcome around 80 on the issue
scale (figure 4.1). This was not Kim's point of view, it was not Bush's point
of view, and it also was not the weighted mean power's point of view. That
mean power perspective, located at the position labeled "Some Reduction / U.S. Recognition," actually had the smallest bloc of power behind it.
Although not supported by many, it was nevertheless the position around
which an initial compromise might most easily have been constructed.
That means that in 2004 one of the first-cut predictions included the
prospect that North Korea could be induced to reduce substantially, but
not eliminate, its nuclear capability in exchange for significant U.S. concessions, including perhaps even diplomatic recognition and certainly including consequential foreign aid (not shown on the figure).

Of course, this forecasting method doesn't tells us how to get Kim Jong
Il and President Bush to agree to the position it represents or to some other
result that could be successfully negotiated. That's where game theory
comes in. Still, it is interesting to realize that, except for working out how to
get Kim and Bush to agree to this result, we now have, with one of our baseline forecasts (and the forecast that ultimately came out of the computer-
simulated game), a prediction, made in 2004, that is very close to the actual
deal struck between the United States and North Korea in 2007.

Let's think now about the game theory side to see the logic that produces a successful result and to get at some of the nuances of that result.
For starters, we should notice that although North Korea is often portrayed in the media as a closed society, a mysterious place about which we
know very little, the truth is that there are plenty of excellent experts in
universities and elsewhere who know a lot about North Korea's government and its leaders. I interviewed three experts, two together and one
separately, and they came up with quite reliable information even though
they disagreed about what they thought would happen in the so-called
six-nation talks. As I intimated earlier, getting the data is not that hard.
The more difficult question is how to frame the issue so that we answer
the right question.

A good approach to solving problems such as presented by North
Korea's nuclear program begins by asking why that country's leaders
would want to develop nuclear weapons in the first place. We can think of
that as asking, "What are they really demanding in the international

arena?" The superficial answer might be to say that they want to threaten the Republic of Korea or American interests in the Korean peninsula. That may have merit but is unlikely to provide the whole answer, or even the main answer. I believe a good place to examine any provocative policy is to ask how that policy affects the prospects of political survival for the incumbent leadership that is engaging in provocation. Remember, this is human nature we're talking about.

Kim Jong Il knows what it takes to retain the loyalty of his military leaders and to keep rivals at bay for years on end. He knows whom to antagonize and whom to placate. He certainly understands that he couldn't defeat a concerted effort by the United States or South Korean governments to overrun North Korea and overthrow his government. But he also understands that he could raise the price of an invasion sufficiently that such an effort would be too costly to consider. In that case, by building a nuclear weapons capability he diminishes or even eliminates the most significant foreign threat to his political survival, freeing himself to concentrate on managing relations with his military leaders, party elites, family members, and senior civil servants. They, of course, could form various coalitions that might threaten his hold on power. Keeping them happy *must* be a primary concern for Kim Jong Il.

If personal political survival is Kim's main concern—and I believe that this is every leader's top priority—then it's likely that there never was an intractable contradiction between his interest in sustaining himself in power and the United States' interest in eliminating any nuclear threat from North Korea. That means all that we had to do was find a self-enforcing approach that provides real commitments on both sides and advances both the elimination of North Korea's nuclear threat and external threats to Kim Jong Il's political survival. That was the strategic problem as I saw it in 2004. My investigations painted a picture of Kim Jong Il as an astute politician whose primary interest in nuclear weapons was as a lifeline to staying in power. Looking at several simulations of his strategic interplay with other powerful North Koreans also made evident to me that he is less monolithic a leader than is sometimes thought in the United States and that he was, and remains, more open to compromise than is generally assumed. What, in 2004, did I contend such a compromise might look like?

I concluded that his demand really was "Assure my security against a foreign invasion." Therefore, the counterdemand had to relate to assuring us that his nuclear threat became moot in exchange for our ensuring his political survival. This meant turning attention away from just threatening Kim to finding a way to make his interests and the international community's interests compatible. Thinking through his interests as well as the interests of others, it becomes clear that any successful compromise requires that the international community be assured that as long as it does nothing to jeopardize Kim's political survival, he will do nothing to jeopardize peace on the Korean peninsula or beyond.

In practical terms this meant that the United States directly, or through third parties, needed to guarantee, and I mean really guarantee, not to invade North Korea. The United States also needed to guarantee a sufficient flow of money—we will call it foreign aid—so that Kim Jong Il's key domestic backers would be assured of receiving substantial personal, private rewards from him. These rewards include money that could go to their secret bank accounts in return for their political loyalty to him. In return for his assured security and for a steady flow of money, the North Korean regime needed to provide a verifiable means of ensuring that its nuclear weapons program stopped. Assurances of his security would most likely come in two forms: formal, explicit Chinese guarantees to defend North Korea, and public American promises not to attack it. Making these guarantees public is critical, because secret assurances are just cheap talk. They are easily violated without imposing political costs on the guarantor who reneges. As for the assurance of a steady flow of money, we are probably talking about as much as $1 billion per year for as long as Kim Jong Il's regime survives. That may sound like a lot, but think about how much was spent every day in Iraq for years on end, and at what additional price in American, allied, and Iraqi lives, and you'll see that $1 billion a year is small potatoes.

There is plenty about such a deal that is distasteful. Bankrolling such a horrible human being is no happy task. It would be ever so much more satisfying if we could just persuade him to do the right thing, but then Kim Jong Il wouldn't be Kim Jong Il if that were feasible. Exactly because he is so horrible, it is important to figure out how to keep him from unleashing a nuclear war in a fit of pique or fear or resignation that he is through anyway and so has nothing to lose. Remember, my task was to find what would work. The desirability of making it work is what we elect leaders to decide.

I want to emphasize here that I have said "a verifiable means of ensuring that its nuclear weapons program stopped." I have not suggested that North Korea's nuclear weapons program and enrichment facilities be eliminated (positions 80 and above on the issue scale in figure 4.1) or dismantled. This is of fundamental strategic importance in making an agreement credible from both sides—so let me explain what I had in mind.

In the event that the United States or the other participants in negotiations with North Korea insisted on dismantling that country's nuclear capabilities, I believe agreement would have been impossible or would prove short-lived, leading to inevitable cheating. Dismemberment of North Korea's nuclear capabilities as an approach overlooks the essence of Kim Jong Il's interests. It fails to put ourselves, as any good strategic thinker must do, in his shoes, looking at the world from his (perhaps elevated) perspective.

If he dismantled his nuclear capabilities, he would no longer have a credible threat of restarting his weapons program quickly in the event the international community—especially the United States—reneged on its promises. In such an environment we can be confident that the international community would renege, and, anticipating that, he would never allow his weapons program to be dismantled. After all, once his nuclear threat was completely dismantled, hardly anyone would have any remaining reason to follow through on payments to him, payments that help him keep domestic as well as foreign rivals under control. The international community would have even weaker reasons to leave him in power. They would rather replace him with someone more amenable to their wishes once his threat to use nuclear weapons to defend his regime was no longer meaningful. Remember, game theory takes a dim view of human nature, and that includes our nature as well as his. However high-minded we think we are, we would have scant incentive to continue to pay Kim Jong Il to behave well once his ability to behave really badly was eliminated.

Thus, as long as his nuclear program is stopped, disabled, placed in mothballs, with inspectors on site at all times, and not dismantled, he has the ability to restart it if the United States or others renege on payments or security guarantees. Conversely, as long as the United States and others do not renege on payments and security guarantees, he has no incentive to throw out the inspectors and restart his nuclear program. It will have achieved its purpose of giving him a life preserver. His likelihood of extracting a higher price in the future by throwing out the inspectors must be balanced against the realization that if he proves untrustworthy there are alternative solutions to the problems he represents—alternatives he wants to avoid. These alternatives, of course, would need to be and could be acted on (and here is where U.S. interests in acting and South Korean interests in avoiding the possible consequences of such action differ enormously) before he recommissioned his nuclear capability.

So we can see that money and security guarantees for disabling, but not dismantling, Kim Jong Il's nuclear program create a self-enforcing mechanism. Neither side would have an incentive to renege on its part of the bargain. It is a deal that reinforces each side's interests. It is the deal that was struck and that, with minor tweaking and continuous jockeying for position, is likely to succeed.

So that is how simple it is to negotiate a nuclear arms agreement! Just find experts, collect the necessary information, use the computer to discover the impediments to the desired outcome, and work out how to neutralize the impact of those impediments by finding actions that serve the interests of the rival parties. With the help of the computer, we were able to see that neither George Bush nor Kim Jong Il was going to budge on his own. Since they were the two decision makers with the authority to say no to any deal, we had to find a way to overcome their intransigence. To do so, we used the computer model to test the likely effects of alternative degrees of American concessions to evaluate their likely influence on Kim Jong Il's approach to the negotiations. We found that security guarantees, especially a mix of assurances from the United States (not to attack) and from China (to defend North Korea if necessary), coupled with significant economic assistance (approaching $1 billion or so per year) to North Korea, would induce Kim to mothball his nuclear capability and allow continuous on-site inspections and securing of his nuclear facilities. Thus, we saw that we could move him into the range of 60 to 65 on the issue scale, and that George Bush could accept this compromise as well, and—poof!—using the me-first principles of game theory we examined in Chapters 2 and 3, we saw the path to a settlement that could indeed be both predicted and engineered.

We've seen now that it's possible to employ this science in nudging North Korea to alternative strategies. But how does this tool really work in the rest of the infinite world of human interaction and conflict? How must we change our base thinking to work out any variety of predictions and, especially, to change the future?

It is indeed my claim (and part of my livelihood) that outcomes of very big problems with many, many players can be systematically predicted and engineered, but, as you may have noticed, it all starts with asking the right questions. The next chapter will examine what those questions might look like across a wide variety of issues from the worlds of business and foreign affairs.

5

■

NAPKINS FOR PEACE:

DEFINING THE QUESTION

EVERY AFTERNOON AT 3:30 the Hoover Institution's senior fellows get together in the Commons Room for coffee or tea and some of the best cookies on the West Coast. In the summer of 1987, as on many other occasions, I took a break from my research to enjoy a cup of coffee, a large chocolate chip cookie, and a bit of collegial camaraderie. On this particular July day, a distinguished Israeli sociologist, Shmuel Eisenstadt, was visiting. He asked me what I was working on, the standard conversation opener in the Commons Room, and I explained that I was trying to improve my original forecasting model which was then several years old.

Shmuel asked, "So, you can predict how to make peace in the Middle East?" This being a tall order, I responded cautiously that perhaps I could predict what steps might be taken over the next few years to advance the prospects of peace. I emphasized that this required data, not a crystal ball. He asked what data I needed and then, pen in hand, wrote on my coffee-stained napkin and his napkin as well, listing stakeholders, their potential influence, salience, and positions on a scale measuring possible concessions in the context of a multination peace conference. (The Soviet Union was calling for such a conference in 1987.)

Eisenstadt's question to me is pretty close in form to the questions I'm asked to examine—whether by companies pursuing mergers, the Defense Department trying to evaluate a terrorist threat, law firms trying to sort out suits, or the CIA trying to understand Iran's nuclear ambitions—and almost always, the question asked is not actually the question for which an answer really is wanted or needed.

WHAT'S THE PROBLEM?

The big question—how to make peace in the Middle East—is best answered by working out what the many smaller questions are that, taken together, add up to a solution. Framing the problem is usually the hardest part of the prediction and engineering process. I never cease to be surprised that even when billions of dollars or thousands of lives are at stake, decision makers rarely work through what it is they actually need to know. They are usually astonished—fortunately, pleasantly so—to realize that they have not thought systematically enough about their own problem to know what it is they need to know or do.

For me, answering a question like "How can we make peace in the Middle East?" involves breaking this question down to specific issues, to specific choices that must be made. Therein lies the key. Questions need to be about actual choices confronting decision makers rather than about abstract ideas like winning or getting ahead. If you want to win, you need to know what it is that signifies you won, you lost, or you did acceptably well, and by how much. Only after we have drawn out exactly what subjects or questions need precise decisions can we start putting the analyses of all of the issues together so that we might successfully identify the underlying stumbling blocks and find ways around them.

For some problems, framing the problem is relatively easy. Much litigation, for instance, revolves around a settlement price. What a defendant really wants to know is "How much do I have to pay to resolve this lawsuit?" To answer that question, it's just necessary to know what demands players currently say they support, how salient the price is for each stakeholder, and how much clout each could bring to bear.

Of course, even lawsuits are sometimes more complicated than just working out a settlement price. Discussing a problem with a client, I might learn that they also need to know whether all charges can be resolved with one global settlement price or whether the price paid for one part of a lawsuit will influence the price that has to be paid for another part. Once in a while, it gets still more complicated. There might be multiple defendants, so it's important to work out not only how much has to be paid to get the plaintiffs to settle, but how the cost will be distributed across the defendants, their insurers, and other involved parties. This sometimes leads to questions about whether it makes sense to drag in other parties as co-defendants or not. Adding co-defendants involves trade-offs. On the one hand, there will be more potential payers with more defendants. That can be a good thing, of course. On the other hand, more defendants means having to coordinate a strategy with many more businesspeople, their lawyers, and their insurers. That can get confusing and expensive. As you can see, a simple question like "What will it take to settle this lawsuit?" can quickly grow into a complicated, interlaced set of issues in which resolving one question changes how others can be resolved. That means we not only have to figure out what the actual questions are that must be decided, but also the sequence of agreements that leads to the optimal outcome.

Business mergers are still more complex most of the time. Unlike litigation, mergers rarely hinge just on the price to be paid for putting companies together. Lots of other issues enter into a successful merger. But mergers can also be simpler than litigation in that the range of likely merger issues is usually well defined and tends to be pretty much the same across all mergers and acquisitions. The price to be paid (or to be obtained) is always a question, of course, but merger discussions rarely get anywhere if there is a substantial disagreement about the worth of a property. That means that the price rarely is the reason an attempted merger or acquisition fails. Remember the example of a merger between a French and a German bank in an earlier chapter. That deal hinged on whether the German executives had to move; they were willing to take a lower price for their bank if they could stay put in Heidelberg.

Several years ago I worked with an international team of analysts on a major effort to create a unified defense industry across Europe, one that could compete with the American defense industry. We looked at virtually every possible combination of defense firms in Britain, France, and Germany, as well as several possible combinations that would have included defense firms in Italy, Spain, or Sweden. The question put to us was "Can we do these mergers?" To answer that question, my team needed to examine approximately seven different issues for each possible combination of firms for a total of more than seventy individual issues, each representing a decision that could make or break a multibillion-dollar deal. Some of the make-or-break questions, the issues on which agreement had to be reached, included (1) the price, (2) allocation of management control between merged units, (3) the scope of businesses to be included or excluded from the merged entity, (4) employment guarantees for workers in various units across national boundaries, (5) government's role in regulating or sharing in ownership and management of the newly created firm, (6) the timing of the transition to combined working units and teams with shared technology, and (7) where the senior managers would be expected to live.

Merger efforts are more likely to fail because of the "lesser" issues than because of a price disagreement. Yet few executives seem to recognize this when they initiate the process. As a result, they spend millions of dollars on getting good financial advice to sort out the right price to offer, only to have many prospective deals fall through because too little effort has been invested in working out other issues. Sometimes these "other issues" can seem so absurdly small that they are ignored, only to end up being, in hindsight, the deal breakers.

A few years ago I worked on a failed merger in the pharmaceutical industry. The prospective deal was expected to produce great efficiency gains that could have dramatically benefited the pharmaceutical market and prescription drug consumers. Almost all of the executives in the two firms involved were most enthusiastic about the opportunity, with "almost all" being the key phrase. The killer issue that did the deal in involved allocating management control between the CEOs in the two firms. That, of course, is no small question. The absurdity came in how the deal was killed.

The two chief executives hated each other, and, as is the case with many big companies in Europe, there were bitter family issues lurking in the background as well. So deep was the animosity between the two CEOs that everyone agreed a way had to be found for them to work together before going forward. A dinner was arranged for the two of them, and senior aides, who should have smelled a rat, also attended. It took a huge effort just to get this pair to sit at the same table.

The host CEO finally agreed to have the dinner at his home and then, without my or any of his own aides' realizing it, he planned a menu of course after course of the other chief executive's most disliked foods. Amazing as it may seem, this prospective multibillion-dollar deal fell apart over a menu, or, more accurately, over a deeply personal conflict that made all other efforts and points moot.

Working out what the right issues are takes patience, good listening skills, and the ability to steer a conversation toward what will really drive results rather than the inchoate musings of the decision makers. Fortunately, the dinner menu rarely plays such a prominent part in negotiations. For me, an "issue" is any specific question for which different individuals, organized groups, or informal interested parties have different preferences regarding the outcome, and for which it is true that an overall agreement cannot be reached unless at least a key set of players come to agreement on the question. It helps to know whether there is a status quo, and if there is, to make sure that the issue is not constructed to be biased against the existing situation.

Recently I taught a seminar in which my students were asked to pick a world crisis that interested them and to model a way to solve the problem. One group of students decided to examine carbon dioxide emissions (a topic I return to in the last chapter). They defined an issue in which one end of the scale reflected current CO_2 emissions and the other end the stiffest reduction advocated by any environmental group. Everything in between represented possible agreements among the players. Do you see the problem here? I asked if they thought there were no energy companies or others who felt that CO_2 emissions should be *less* regulated than was the case at the time. Of course there were interested parties who wanted to have more freedom to produce CO_2! So the way the students had designed their issue was biased. The only answer that could come out of their exercise was to increase regulatory controls or, with low odds, keep them the same. Their scale allowed no possibility of anyone wanting to produce *more* carbon dioxide. They had unwittingly created a biased issue, one that made them feel good but would almost certainly lead to a wrong answer. Such fundamental flaws in the design of questions ensure that the answers to them will be useless.

YOU CAN'T ALWAYS GET WHAT YOU WANT, BUT IF YOU TRY . . .

Once an issue is properly framed, we have to think about how to capture the thought process that people go through in working out decisions. Without doing that, without climbing into the heads of your rivals, you're not likely to get what you want. You're not even likely to know how to try to get what you need.

The game structure I use looks at choices as sometimes involving co-operation, sometimes competition, and sometimes coercion. The most complicated part is to try to emulate how people think about changes in their situation as well as what others say and do. Players are always interested in altering the lay of the land in their favor. They want to surround their desired outcome with tall mountains of power that are hard to overcome. They want to tear down mountains of opposition, leveling the power terrain around positions they want to defeat. At every step along the way, everybody has to work out who will help them and who will get in the way. They have to calculate the risks and rewards, costs and benefits of actively trying to change other people's choices or lying low, trying to stay out of the line of fire. The math can get complicated,[1] but let's look at some examples of the process at work that we should be able to follow pretty easily.

The table below shows the small data set that resulted from the cookie-and-coffee conversation I had with Shmuel Eisenstadt in 1987, augmented in 1989 by a discussion with Harold Saunders, who by then was the former deputy assistant secretary of state for the Near East and North Africa. The continuum of possible outcomes on the settlement issue ranged from the establishment of a fully independent, secular Palestinian state at one extreme to the annexation by Israel of the West Bank and Gaza at the other extreme. Position 30 on the scale was defined to represent territorial concessions granted by Israel to the Palestinians without establishing an autonomous state but establishing instead a Palestinian political entity federated with Jordan. The 1987 status quo was located at the position equivalent to 85 on this scale (under the "Negotiated Settlement Options" column). There was no semi-autonomous Palestinian territory or government at that time.

Player	Influence	Negotiated Settlement Options	Salience
Israeli Settlers	100	100: Israel annexes West Bank and Gaza	99
SHA	85	85: 1987 Status Quo	85
Hard-line Likud	85	85: 1987 Status Quo	90
Likud	60	70: Palestinian territory with weakest autonomy	50
Israeli Defense Forces	100	60: Semi-autonomous Palestinian territory	80
Labor Party	60	30: Close federation with Jordan	75
OCC	100	25: Close federation with Jordan	95
PEA	70	20: Loose federation with Jordan	85
PLO	100	20: Loose federation with Jordan	95
PFLP	20	0: Independent, secular Palestinian state	95
FND	10	0: Independent, secular Palestinian state	80

It is not hard to see how territorial concessions can be organized on a continuum. Although the scale above is not based on percentages of land or land value, still there is a natural progression in choices ranging from Israel's annexing contested territory on to no concessions by the Israelis and finishing at the other end of the scale with a fully independent Palestinian state. From these beginnings in 1987, I prepared a forecast that would closely predict the actual territorial concessions agreed to between the Israelis and the Palestinians in 1993 at Oslo. (We will look more closely at this forecast a little later in this chapter.) Let me reemphasize a central assumption behind the scale, an assumption I introduced casually in the discussion of North Korea. The scale tells us that someone advocating a weakly autonomous Palestinian territory (70 on the scale) strongly prefers the status quo (at 85 on the scale) to a territory closely in federation with Jordan (25 on the scale). We know the preference for 85 is stronger than for 25, even though 85 and 25 flank the advocated position at 70, because 70 is much closer numerically to 85 than to 25. That is how each scale works. Numerical values that are closer to the numerical value of the player's advocated position are liked better by the player than positions reflected by numerical values farther from the advocated position. Knowing that, let's have a look at a business issue that can help us understand more about how to turn problems into a well-defined numerical scale.

■   ■   ■

Here the focus is on what an issue's scale might look like when the question involves something less obviously numerical in character than land for peace. To see an answer to this, consider the following range of choices from a litigation I worked on some time ago. Naturally I have masked the details to protect my client, but the ideas should be clear enough.

WHAT CHARGES WILL THE U.S. ATTORNEY BRING AGAINST THE DEFENDANT?

Scale Position	Meaning of the Numerical Value on the Scale
100	Multiple felony charges including several specific, severe felonies
90	Several specific, severe felony counts but no lesser felonies
80	One count of the severe felony plus several lesser felonies
75	One count of the severe felony but no other felonies
60	Multiple felony counts but none of the severe felonies
40	Multiple misdemeanor counts plus one lesser felony
25	Multiple misdemeanor counts and no felonies
0	One misdemeanor count

This scale makes clear how the client prioritized possible outcomes. The definition of 0 on the scale tells us that the client did not believe that any stakeholder would argue for the dismissal of all charges. That was not a choice included in the range of feasible outcomes. The upper bound on the scale indicates that the client believed at least some in the U.S. Attorney's Office or others involved in the process would argue for severe criminal penalties. The real action was to unfold in between, and we'll take a deeper look at this "case" in a later chapter.

Let's take a quick look at this issue scale. It can shed a lot of light on how to think about issues. We can see that the two key "distance" differences here, laid out by the client, are the plea from 25 to 40 (no felony vs. one lesser) and from 40 to 75 (no severe vs. one severe). After that, life gets worse, but not nearly as dramatically worse as happens with the move from 40 to 75. This tells us, among other things, that avoiding a severe felony charge altogether was of greater value to the client than adjustments in the number of severe felonies that might have to be pleaded to.

In framing issues this carefully, the decision makers begin to learn things they had not recognized about their own problem. Instead of an amorphous question, they now have focused issues to address. They have confronted the problem and defined what the meaning is behind the choices they face. Once they go through the interview process to identify the remainder of the data—who the players are, where they currently stand on the issue, how salient it is to them, and how much clout they can each exert—they have, for the first time, a genuinely comprehensive view of how high a mountain they must climb. As I mentioned, we will revisit this case in a later chapter to see how they thought it would turn out and how it actually did turn out. For now, let's return to the napkins covered with data on how to promote peace in the Middle East.

PEACE IN THE MIDDLE EAST?

The information scribbled on a couple of napkins turned out to provide a terrific grounding for figuring out what was going to happen in the Middle East over the several years following 1987. Eisenstadt knew his stuff, and so did Harold Saunders. What did we learn when the data were subjected to the model's dynamics, allowing all of the players the opportunity to bargain with each other, trading territorial concessions for political credit? The predicted outcome was reported in a 1990 journal article I wrote based on the cookie-hour napkins.[2] The article was published three years before the Oslo Accords created the Palestinian Authority. It predicted resolution at position 60 on the issue scale out of a feasible range of 0 to 100. Sixty on the scale was defined at the time as being equivalent to the Palestinians' gaining some weak localized territorial control over a semi-autonomous entity. Remember, the status quo was 85, and lower values reflected greater and greater concessions to the Palestinians. We now know that the territorial concessions that were actually worked out between the Israeli government and Yasser Arafat were equivalent to about 60 on the scale, as I predicted in print three years beforehand.

Of course, this prediction was made a long time ago and was intended to be good only once there was a change in Israel's government. As I wrote at the time, "Given [then Israel's prime minister Yitzhak] Shamir's apparent perception of the situation, we must conclude that as long as he is Prime Minister of Israel, there is no reason to expect significant progress." Shamir, who became Israel's prime minister in October 1986, was replaced by Shimon Peres in July 1992, paving the way for real progress between the Israelis and the Palestinians.

The analysis, however, was more detailed and nuanced than just a prediction of territorial concessions. My 1990 study went on to suggest that the then large and influential Popular Front for the Liberation of Palestine (PFLP) would "become politically isolated and irrelevant in negotiations. We can anticipate that they would respond to such a situation by increasing terrorist acts, aimed not only at the Israelis but perhaps also at the PLO leadership. If the analysis is correct and if the PLO adopted a strategy of incremental moderation, the future of the PFLP would be to flicker out of the picture." Remember, this was out there for anyone to see and read in 1990. That is what I meant earlier when I said we must be willing to risk embarrassment if we want people to have confidence in our predictions. Certainly few in 1990 would have argued that the PFLP was likely to "flicker out of the picture."

Today, two decades later, we can look back to see how accurate that prediction was. For instance, what do independent sources now say about what happened to the PFLP following the creation of the Palestinian Authority in 1993? BBC News Online, reporting on January 16, 2002, quotes Abdel Bari Atwan, editor in chief of the London-based Arab newspaper *Al-Quds*, as saying "The movement [i.e., the Popular Front for the

Liberation of Palestine] had become marginalized. . . . It had gone from being the second most important Palestinian group to forth [sic] or fifth."[3] Anthony Cordesman, a highly regarded Middle East scholar and expert commentator for ABC News, echoes the BBC view. Reflecting back on the PFLP, he writes, "The PFLP opposed the Oslo peace process. As the PA [Palestinian Authority] and Arafat's Fatah gained strength, the PFLP became increasingly marginalized."[4] Similarly, GlobalSecurity.org, generally seen as a reliable source of information, notes that, "Once a key player in Palestinian politics, the PFLP lost influence in the 1990s and was sidelined as Yasir Arafat established the Palestinian Authority." The PFLP's decline is similarly reported by numerous other sources. Most of these sources attribute the PFLP's decline to events that happened after my 1990 article was published and long after Shmuel Eisenstadt wrote numbers on a couple of napkins. That is, the modeling exercise undertaken before the 1991 Gulf War, before the first intifada, and before the Oslo Accords foresaw the territorial changes between the Israelis and Palestinians, the necessity of a Labor government coming to power in Israel, and the decline of the PFLP. It foresaw the fundamental developments far enough ahead that even today people have trouble recognizing that the seeds of the agreement in 1993 had already been planted and were growing, unseen by most onlookers, well before.

What was it that the model identified in the give-and-take between the Palestinian side and the Israeli side based on data collected in 1987 and 1989 that led to a successful prediction? This is really the crux of the matter. After all, working out that the predicted outcome could lead to a stable agreement involved more than just pondering Eisenstadt's original, vague question: "So, you can predict how to make peace in the Middle East?" Since Arafat and the Israeli prime minister both had vetoes over any deal, it was essential to figure out whether they were likely to end up supporting the same agreement, and if so, why. Here's what the model indicated:

Israel's Labor Party leader, Shimon Peres, was not in power when the study was done, and it was evident in the model's results that until Labor came to power no agreement could be reached. That meant that the focus of attention in analyzing the possibility of a peace deal—the problem put to me by Shmuel Eisenstadt—had to be on Peres and the Labor Party.

The model's logic produced output that indicated that Peres believed he would face a lot of political pressure at home over his stance regarding the Palestinians. To attain and retain power—the goal of every politician—he believed (according to the model's logic) that he had to look as though he would be tougher in Palestinian negotiations than was implied by the quite moderate stance he took in the late 1980s. The model showed he needed to be only a bit more moderate than Shamir, leading him to agree to occupy a position in the mid- to upper 60s instead of Shamir's drift between 70 and 85 on the scale. So Peres was predicted to be responsive to political expediency at home.

For his part, Arafat concluded, according to the model's simulations, that he needed to moderate his own stance to keep Labor from taking too hard-nosed a view within Israel. The model suggested that he would choose a course of action based on his personal political welfare rather than the well-being of the Palestinian people per se. As I wrote in the 1990 article, "The model solution suggests that Arafat would stabilize his political position, leaving himself devoid of serious political opposition either among the Palestinians or within Israel. . . . If Arafat does choose to moderate his stance, this suggests that he is willing to sacrifice both the Palestinian cause and his opponents at the altar of his personal political welfare." That seems to have been the case. Thus the analysis went from a vague question to precisely structured propositions and a detailed analysis of the negotiating dynamics that made it impossible for Shamir and Arafat to reach agreement and those that made it possible for Arafat and Peres to come to terms once Peres became prime minister.

Asking the right questions and isolating the key interests for a given problem is too often a step that's not taken from the beginning. Instead, we settle for conventional wisdom about the reasons behind actions that seem to fit the puzzle. The costs of this laziness can be grave, particularly when the problems are the kind for which society seeks remedies in order to prevent their recurrence. With an initial misdiagnosis, the wrong treatment will most assuredly follow. Had policy makers paid attention sooner to the pulls and tugs likely to face Arafat as well as the Israelis subsequent to Oslo, perhaps they could have managed circumstances better and might have avoided many of the setbacks between the Israelis and Palestinians since 1993.

DON'T JUST LOOK WHERE THE LAMPPOST SHINES

The failure to zoom in on what the issues are, who the players are, what their incentives are, and how to fix those incentives is not limited to problems in foreign affairs. The business world suffers at least as much from the same difficulties. To take an example of this from recent years, let's look at a model that colleagues and I developed to identify the causes of and solutions to corporate fraud.

Since the Enron debacle, Congress has made a real effort to strengthen corporate incentives to report honestly by introducing a massive amount of new regulation through the Sarbanes-Oxley bill, passed in 2002. However, I'm afraid Sarbanes-Oxley touches on but does not nail the root causes of fraud, at least not as those causes are seen by the model my colleagues and I developed. As we can see in figure 5.1, fraud litigation is once again on the rise, despite the passage of Sarbanes-Oxley. Since the recession began in 2007, fraud has skyrocketed ahead of its pre-downturn

FIG. 5.1. Federal Securities Class Action Litigation, 2002–2008

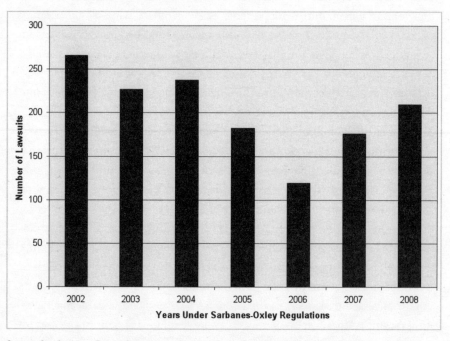

Source: Stanford Law School Class Action Clearing House, http://securities.stanford.edu/

2006 level. This is just a bit of evidence—there is more to come—for my model-based belief that the premises that guided Congress in passing Sarbanes-Oxley were misguided. President Obama promises a new wave of regulatory controls to rein in the risk of business fraud and failure. Let's hope that his administration and the Congress are more attentive to incentives so that they get the regulations right.

To regulate the risk of fraud it is fundamental that we first understand the motives for committing fraud. How does the model that my colleagues and I worked on determine which companies have an incentive to commit fraud and which do not?

Our game-theory approach is a variant on the study we did to understand how nations are governed (recall dear old Leopold). But pay attention, as the context of the corporate setting provides a most interesting twist. As we saw in our study of Leopold and other heads of state, loyalty to leaders is much weaker in democracies because the competition is over policy ideas rather than the personal enrichment of a few supporters. Much the same might be said for corporations and the survival of corporate executives. From this starting point, we might assume that the outcome in the business world would be the same as that in the political: "autocratic" leadership leads to corruption (fraud), while more "democratic" leadership does not. *This is not the case.* Perversely, as we will see, the strong loyalty engendered by relatively autocratic corporate styles helps *reduce* the risk of fraud. To understand why, we have to leave the light of the lamppost and really look behind the scenes at the logic that governs corporate behavior.

Big firms have millions of shareholders. Yet few of them attend the annual shareholders meeting, and they have little idea of how the business is run or how it might have done if it had been run differently. They dutifully send in their proxy, voting the way the board suggests, or they toss their proxy statement in the recycling bin and do not vote at all. Big blocs of votes are controlled by a small number of institutional investors, senior executives, and directors. They, not the shareholders, decide how to run the company. The more institutional investors and powerful shareholders there are, the more parties to whom the management is accountable. Ring a bell? It's the challenge presented to any leader: the more "democratic" the system (think of democracy not as an absolute concept but rather as a continuum), the more people to please.

When things are going well, the incentives for the executives running a company are not incompatible with the shareholders' interests. Growth in profits is good for the executives and it is good for the shareholders. But sometimes things do not go well. Then the interests of management and shareholders might part ways. Let's see why.

In developed equity markets, the fraud model indicates that accounting fraud typically results because management is trying to preserve shareholder value. Don't get me wrong. The fraud model does not think of executives as altruists who lose sleep trying to think up ways to make shareholders better off. They commit fraud to protect their jobs in the face of poor performance rather than as a result of a desire to defraud investors per se. That means we can use public records to link the likelihood of fraud to any publicly traded corporation's reported performance, ownership oversight, and governance-induced incentives to manage the firm truthfully.

Examining publicly available information taken from SEC filings, my colleagues and I found that the amount paid in dividends to shareholders and salaries to senior management during years of honest reporting and during years immediately preceding fraudulent reporting differ in significant ways. In the fraud years, senior management receive less compensation— you read that right, *less* compensation, not more—than expected given their corporate governance structure and reported corporate performance. Dividends typically also fall short of expectations. All the while, reported performance and therefore the firm's growth in market capitalization look healthy, just as they do in honest years. It is important to note that the compensation for senior management may still have increased in such fraud years, and in an absolute sense may still be quite grand, but the critical point is that if things are really going so swimmingly for the company, then compensation should be even higher than it is.

The logic behind the model we designed pinpointed these as trends to look for. We didn't know whether we would actually find these patterns in SEC filings, but we did know that they were the key to predicting fraud if our model was right.

Why these patterns? CEOs always have incentives to take actions that protect their jobs. If they see that their company is not performing up to market expectation, they are at risk and will take action to salvage their situation. Now, they can argue that the firm is the victim of unforeseeable shocks for which they should not be held accountable (this was the argument made by the CEOs of GM and Chrysler in seeking a government bailout—they pointed to the economic downturn, not management's decisions, as the cause of the auto industry's woes), but such arguments are risky to say the least, and may not be adequate to protect a CEO's job.

If blaming the economy or some outside force doesn't salvage senior management's situation, then the top executives might misrepresent the corporation's true performance. If they can sell the belief that the company is performing just fine, then they won't be at risk of being fired. It is difficult for outsiders to know the corporation's true volume of sales, revenue, costs, and profits. Market capitalization reflects these factors, and indeed these are the factors that when falsely reported and subsequently detected result in accusations of accounting fraud.

If revenues are exaggerated or costs are understated, then senior executives can temporarily lead the marketplace to misjudge the true worth of a company, making the company appear (falsely) to have met or exceeded expectations. This, the model suggests, is the essential motivation behind corporate fraud.

This wedge between lower-than-expected stock dividends and compensation for executives and seemingly normal or good growth in market capitalization is therefore an early-warning indicator of an elevated risk of fraud. Neither the SEC nor many corporations, however, seem to realize the importance of this information in detecting early signs of trouble. Sarbanes-Oxley certainly does not draw attention to analyzing the size of this benefits wedge.

The game's logic and the evidence culled from more than a decade of corporate filings across hundreds of firms also raise questions about journalistic accounts and popular perceptions. A pretty standard journalistic view of fraud is that greedy executives act to enrich themselves at the expense of shareholders and employees, that they are little more than looters, and that the problem boils down to outrageous character flaws. This kind of thinking gets us nowhere.

Too often we look at what happened most recently and assume that earlier actions were motivated by those ends. It is easy to believe that greedy executives cook the books to enrich themselves, with self-interest tied merely to short-term gain (of course I would never argue that self-interest isn't the key motivation, but we must examine exactly what the nature of that self-interest is). But in doing so we forget, for example, that Enron's fraud started around 1997 and yet the senior managers did not sell off their shares until around 2000 and 2001. Why would they have waited so long, risking discovery for years, before cashing out? True, stock prices were going up, but equally true, the risk of being uncovered grew greater and greater with each passing month and year. Did they really only seek the gains to be made in the period from 1997 to 2001, or were they hoping for much greater gains five, ten, or fifteen years on?

We won't analyze the problem properly if we look for the causes of fraud in its end result. Remember, correlation is not causation—the beginning, not the end, is where the explanation lies.

In the Enron case, it seems clear that by 2000 or 2001 the most senior leadership in the company realized that they could not fix its problems. Having reached the end of the corporate game, they cashed out. It is despicable that they did so while covering up the true state of affairs, thereby leaving their pensioners on the hook. But it seems equally evident that Enron's senior management did not hang on for four years before cashing out just to enrich themselves and walk away. All that time, according to the game's logic, they were trying to save the company because their longer-term interests would be rewarded if they were successful in doing so. If the numbers had to be fudged while they corrected the company's course, so be it. In their view, the end justified the means. None of that was going to happen if the shareholders, and especially if key board members, found out what trouble Enron was in.

Give executives the wrong incentives and you can count on their taking actions with bad social consequences. Give them the right incentives and they will do what is right, not because they are filled with civic virtue but because it will serve their own interests. Remember Leopold? He had pretty good incentives in Belgium and he did good things there. He had horrible incentives in the Congo and he did horrible things there.

What, then, are the right and wrong incentives? Why do some companies commit fraud while others—the vast majority of firms—even in dire circumstances do not? In answering these questions we can gain insight into how to alter incentives appropriately and how to anticipate who has the wrong incentives and is at serious risk of committing fraud.

One clear implication of the fraud model my colleagues and I developed is that the broader the group of people CEOs rely on to keep their jobs, the more likely it is that the shareholders who put them in power will throw them out. That's what happens to leaders of democracies, and that is what is more likely to befall underperforming CEOs in relatively democratic companies. To save themselves, they are perversely incentivized to misrepresent the corporation's true performance so that they don't have to explain underperformance in the first place.

This is not to say that more "autocratic" companies (fewer people to please in the power structure) are incapable of fraud. It's just that things have to be considerably worse for those companies before management sees sufficient risk to their jobs that they are tempted to engage in fraud. Our sliding scale extends across the public/private company divide as we consider partnerships (think of them as oligarchies) and family companies (monarchies).

Government regulators and boards of directors could do a better job of protecting shareholders and employees from the risk of fraud. To do so, the focus needs to be more squarely placed on the incentives executives have to monitor themselves and their colleagues in the face of declining business performance. Knowing how to adjust governance structures to induce the right incentives is the way to regulate firms successfully. Balancing incentives in good times and bad is a major challenge for running a business in a way that attracts and retains top-quality executives and satisfies shareholder expectations. Optimal corporate governance design needs to be done on a case-by-case basis, taking the nature of the firm's market into account. A sweeping regulation cannot facilitate the fine-tuning that is needed to get incentives right. Confidence in business requires that we move in these directions rather than putting our energies into finding greedy individuals to blame or one-size-fits-all fixes for what are manifestly corporate governance problems. Looking for greed is just like the drunkard looking for his keys under a lamppost. More often than not, what's lost is not under the bright lights.

In a broader sense, if we truly want to make it easier for corporate executives to come clean about problems they discover as soon as they discover them, then we also ought to change the law so that they are not punished for spilling the beans on themselves.

Lots of companies discover problems with their products or their performance long before these problems become public. Indeed, it is a good bet that some serious problems never become public at all. A few years ago, for instance, the 3M Corporation pulled Scotchgard off the market. Scotchgard was one of its biggest earners, and yet one day it was in the stores and the next it was gone. A few years later, 3M introduced what it called a new, improved Scotchgard. The EPA and other firms in the chemical industry wondered whether 3M had discovered a health or safety risk associated with the main chemical in Scotchgard, a chemical not found in its "new and improved" product.

I don't know whether 3M discovered a problem or just decided one day to change a successful product. Imagine that they did find a problem. What would they—no, better yet, what *could* they responsibly do? Company leaders in such situations may be eager to reveal whatever it is they've discovered, but they also realize that doing so would violate their fiduciary duty. They are damned if they do and damned if they don't. A public announcement leaves them open to lawsuits by people who used the product before anyone—inside or outside the company—knew there was a problem. These suits can be devastating to shareholder value, and it is shareholder value that corporate directors are legally obliged to protect.

Probably many companies would reveal what they know when they discover trouble if the government would immunize them against prosecution for any problems in their products that were previously unknown to them. The government won't. Litigation is the favored solution, as opposed to rewarding responsible, public-spirited actions by corporate executives in difficult straits. The result is that corporate leaders are given the wrong incentives. Remember all the litigation surrounding the problems caused by DDT? Do you also remember that the Royal Caroline Institute won the Nobel Prize for Physiology or Medicine in 1948 for developing DDT? With litigation run rampant, we fail to provide corporations and their leaders with protection for reasonable expectations and decisions that, not by any misdeeds on their parts, may simply turn out to be wrong.

In this chapter, we've explored how to frame questions. The main idea is to isolate the individual components of a problem that shape its resolution. Then it's a straightforward matter of turning those isolated individual components into issues that, depending on the circumstances, may be decided separately or that may be linked to each other. Once an issue is well defined, experts have an easier time talking about who really will try to influence the decision on each item. Then we can have the computer play the forecasting and engineering game to simulate what proposals each player is expected to make to each other player on a round-by-round basis, and we can bring into relief the incentives that players have to accept or reject proposed solutions.

With the computer program at the ready, we can sort through the problem and not only predict results, as I did with the napkins, but begin to engineer results to change outcomes, as I hinted could be done to prevent corporate fraud. Engineering outcomes is the subject of the next two chapters.

6

■

ENGINEERING THE FUTURE

IPLOMATS ARE CONVINCED that a country's name is an important variable that helps explain behavior. That's why the Department of State is organized around country desks, just as the intelligence community is organized around geographic regions. Leaders of multinational corporations take much the same view. When they have a problem in Kazakhstan they call their guys in Kazakhstan to find out what to do. That seems eminently reasonable and right. Yet it is only partly right and terribly inadequate for solving most problems, or, as I see it, for engineering the future.

Now don't get me wrong. Knowing about places, and how different they may be, is important, but, perhaps surprisingly, it is not as important as knowing about people, and how similar they are, wherever they are. I have not arrived at this view lightly nor, I hope, in ignorance. As a matter of fact, the training that led to my Ph.D. molded me into a South Asia specialist. I even studied Urdu for five years, both during my undergraduate and graduate studies, and did field research in India—so I certainly respect and value area expertise. But I don't think it's the way government or business should organize itself for problem-solving purposes.

Here, as in so many other ways, I am swimming upstream against a strong current. Mine is a controversial stance in many of the circles in which I travel, and many in those circles see views such as mine as foolish at best and dangerous at worst. Still, I do not shy away from the risk of publishing predictions about things that have not happened—and by and large, those who disagree with me do not do the same.

As valuable as area studies is, it is by itself a poor substitute for the marriage of expertise about places and the expertise of applied game theorists about how people decide. Yet we seem to think that knowing the facts is sufficient. Some even contend that it is ridiculous to rely on something as abstract as mathematics to anticipate what people will do. Speaking of ridiculous things, we surely would think it ridiculous if chemists believed that oxygen and hydrogen combine differently in China than they do in the United States, but for some reason we think it entirely sensible to believe that people make choices based on different principles in Timbuktu than in Tipperary (we might be different from mere particles, but we're not all that different from one another). Country expertise is no substitute for understanding the principles that govern human decision making, and it should be subordinate to them, working in tandem to provide nuance as we actively seek to engineer a better future.

To explore how this view informs the predictioneering process, in this chapter we'll turn to a group that lives and feeds on human conflict: lawyers. Lawyers share with diplomats, academics, and business leaders a conviction that country names matter, but—let's give them *some* credit— they believe this for better reasons. Different countries have adopted different rules of law. Some assume innocence until guilt is proven, while others assume the opposite; some make the loser pay the costs of litigation, while others do not. But that difference aside, lawyers, like diplomats and statesmen, spend much of their time negotiating the resolution of disputes—and almost none of them ever learn any game theory or study negotiation strategies. Lawyers study law, and diplomats study countries. Both groups may read popular books from which they get some useful insights, but collections of clever anecdotes and off-the-shelf recipes for success are no substitute for the serious study of game theory or for getting the assistance of people with expertise on how decisions are made.

When diplomacy is successful, wars are fought with words, the combatants sitting around a table, drinking Perrier until a resolution is reached and celebrated with fine wine. Lawsuits are wars too. Just as most international disputes are settled long before they get to the battlefield, so are lawsuits played out in meeting rooms, boardrooms, lawyers' offices, and only rarely courtrooms. Lawyers prepare arguments, investigate precedents, amass documents, and study the other side's paper trail. In big corporate suits, millions, even tens of millions of dollars are spent on armies of attorneys. I've worked on lawsuits involving so many lawyers and so many law firms that it was almost impossible to keep track of them. My consulting company has frequently advised on lawsuits in which the defendant— usually our client is a defendant—holds uncountable meetings with one or two dozen lawyers at each meeting, all senior partners in major law firms, each one billing $400, $500, $600, $700, or more per hour. A typical meeting running just one workday with a dozen lawyers in attendance costs about $57,600—and that doesn't include the vastly higher costs of preparation for that day's discussion. Multiply that by dozens of meetings, and then two or three times more to reflect the costs of preparation, and you begin to see how quickly lawsuits become phenomenally expensive. I've advised on cases where my client spent tens of millions more on lawyers than they paid in settlement.

Of course, all that lawyer money is not spent without some reason. It's generally spent for two purposes. First, the money is buying preparation to improve the prospects of victory. That is, of course, the lawyers' job. Second, the money is spent to signal the other side that they are up against deep pockets that can endure high costs to fight the good fight. The message: "We will keep you embattled in motions, countermotions, and delays until we break the bank. We can spend more than you." Naturally, the other side is spending tons of money with the same two purposes in mind. The two sides are playing the game called the war of attrition.[1] It's great for lawyers, and awful for everyone else.

Armies are the diplomat's analog to the prodigious spending on lawsuits. Having more and bigger guns discourages others from picking fights with the well-armed. Deterrence works much of the time, but sometimes, just as with the deterrent threat of costly litigation, arms fail to protect the peace—and war results. Wars and litigation are inefficient ways to resolve problems. They almost never end in a decisive victory. Instead, they usually end in a negotiated settlement. Both sides find a deal they could, in principle, have arrived at without all the costs that finally brought them to the negotiating table.[2]

One reason that diplomats and attorneys do not avoid the huge costs of their pre-settlement minuets is that they simply don't know bargaining theory. They're reduced to working out the complexities of each situation on their own. Seat-of-the-pants wisdom and experience help, and some lawyers and diplomats are quite good at it, but for most, an awful lot gets overlooked, delaying settlement and rendering the resolution of disputes more costly than it needs to be.

In the dance that precedes negotiations, attorneys and diplomats tend to center their arguments on the merits of the case. Rarely do they really think through the motivations and incentives of their opponents, the people they represent, and themselves. After they form an opinion about the strengths and weaknesses of their case—which is what lawyers are trained to do—they try to impose some reality check on what their clients think, whether the clients are plaintiffs who think themselves terribly harmed or defendants who think themselves exploited. The positions the defendants (or plaintiffs) discuss among themselves in terms of money or other factors in a lawsuit reflect their judgment about the merits of the case. They know this is true for the other side as well. It is no different with international negotiations, whether they concern land for peace, the abandonment of nuclear weapons, or basic principles of governance and human rights.

When my firm advises on a lawsuit, we are always asked how much of the documentation we want to read. Presumably the attorneys hope to give us a reality check, just as they have done for their clients and with themselves. Usually the documentation we are given the opportunity to read is prodigious. There are stacks and stacks of paper. Fortunately, our answer reliably is that we really do not need to read the documents. The merits of the case don't matter very much once negotiations begin—*the merits are inherent in the impasse.*

How can this be? Remember the information we seek in expert interviews. None of it is about how meritorious anyone's position is. It's all about calculating how much they care about the result and ferreting out how much they care about getting personal credit. Business managers often care a lot about the result; lawyers often care a lot about credit that results in getting more business and building their reputations for the next suit.

In the last chapter I introduced one issue that was part of a more complex web of problems confronting a large firm embroiled in litigation with the U.S. Department of Justice. Here was the table I used to reference the scale of outcomes:

Scale Position	Meaning
100	Multiple felony charges including several specific, severe felonies
90	Several specific, severe felony counts but no lesser felonies
80	One count of the severe felony plus several lesser felonies
75	One count of the severe felony but no other felonies
60	Multiple felony counts but none of the severe felonies
40	Multiple misdemeanor counts plus one lesser felony
25	Multiple misdemeanor counts and no felonies
0	One misdemeanor count

Now I'd like to follow the process through in this example to enable you to see how an outcome can be engineered.

The second appendix contains the information on the plea issue as obtained from the defendant's experts, including in-house lawyers, outside lawyers, and corporate executives. Naturally, it is masked to protect anonymity. As is evident from a glance at the appendix, just on this one issue—and there were many others in this litigation—the list of interested parties was extensive. Not only did the defendant have a long list of attorneys and corporate executives with an interest in trying to shape the charges brought against the firm, so too did the community that had been affected by the firm's actions, as well as various segments of the federal government. Far from being unusual, this is typical of a large, potentially costly, and even devastating litigation.

The long list of involved parties means the game was much more complicated than anyone could possibly keep straight in their head. Up to now you may have looked at examples and thought "I can work this out in my head," but no one can work through this complicated a problem without the help of a computer. That is exactly where the added value comes in from having a trustworthy algorithm.

We've all heard stories about evil corporations defrauding people, flouting safety, addicting people to their products, avoiding taxes, polluting the environment, running sweatshops, and God knows what else. That's what this looked like to me at first blush. The company in question, my client, was accused of having done really terrible things that prompted not only civil action but criminal complaints as well. They were accused pretty much of having destroyed a local community for profit. And yet they seemed like such nice, friendly, soft-spoken, genuinely good people. They had pictures of their children and grandchildren in their wallets. They drove modest cars, ate in normal restaurants, and watched the usual run-of-the-mill TV sitcoms and reality shows. Could the situation really have been as awful as it was portrayed? Could the cast of characters hiring my consulting firm really be the soulless monsters described in the media?

As is almost always the case, reality was a lot more nuanced and complicated than the charges suggested, and the people involved were not the satanic ogres portrayed in the local press. Terrible things had happened, but it was far from obvious that the company was responsible, culpable, or negligent or that it harbored the slightest bit of intent to do any harm, for profit or for any other reason. In fact, as dramatic as the news stories were about the case, reality was much different, and it seems—as we will see—that the U.S. attorney, if not his or her office, understood that.

Mind you, I am not trying to justify some of the awful things that happened. My partner and I strongly urged our client to take steps outside the litigation to help the community involved, just out of a sense of concern for the people's well-being. They welcomed our advice and acted on it. They wanted to do something for the people who were harmed and had already been contemplating some such action, even though their attorneys urged them not to, fearing this would be interpreted as an admission of guilt. Our advice just served to tip the balance in favor of their doing what was right over what was expedient. That decision, however, was related to humane behavior, not to justice.

Justice requires that we distinguish between bad outcomes and bad intentions or willful ignorance. I don't believe it's fair to blame people when things turn out badly unless there is proof (and not just innuendo) that they intentionally chose actions or inaction when a reasonable person could foresee the bad consequences of their decision. It is best to judge people based on what they reasonably could know and expect before they did things, not based on what we know later, after the situation has played itself out. But of course I am no lawyer, so my view of justice may be way off compared to how the American judicial system thinks about things. After all, it is not a lawyer's job to get at the truth, it is a lawyer's job to make the best case possible for the client. That, I suppose, is my job too when I'm wearing my consultant's hat instead of my professor's hat.

Anyway, the unfolding drama required a big stage. It involved at least a metaphorical cast of thousands. Still, the final decision process revolved around a few star performers, many of whom were most reluctant to see their names in bright lights. The lead players who longed for anonymity included the board of directors of my client's firm, some of whom were pretty actively engaged in discussions over how to handle the issues; the president and the CEO of the relevant unit of the corporation; and the senior in-house attorney. The senior outside attorneys were also crucial players in the unfolding drama.

On the other side, the U.S. attorney, his/her staff, some of the line attorneys in the Department of Justice and in ABC (a government agency whose name is masked to ensure anonymity for the client), the head of the relevant local government, and plaintiff's counsel didn't mind seeing their names in lights at all; in fact, a few of them relished the thought. They were star players as well.

Getting the lead actors to agree on a settlement was the task at hand. Otherwise, the cases were going to go to trial. Probably the corporation and its representatives would have come out pretty well in terms of a judgment, but not before they had been dragged through the mud day in and day out during the trial. That was the scenario expected by the client. It wasn't a pretty picture. They had been working on the drama's script for several years with no sense of progress but with deep concerns that catastrophe lay just around the corner.

Keenly aware of the aura of doom and gloom that pervaded all discussions, my partner and I set out to define the issues and to turn the model loose in order to get a first impression of the lay of the land. We were curious about whether the situation was as hopeless as the client thought. The model's initial estimate of what would happen—our weighted mean and median are the same in this case—equals 60 on the outcome scale. Sixty is equivalent to pleading guilty to multiple felony counts not including any of the severe felonies. This initial prediction was viewed as good news by the defendant, a ray of sunshine in their overcast view of the situation.

The firm's most senior executives—not just bit players in the company—were looking at having to plead guilty to at least one count of a severe felony as well as several lesser felonies. So the initial estimate revealed the possibility of a better-than-expected outcome for the client. That was the good news. But every silver lining needs a cloud, and this was no exception. The simulation of the negotiating game that followed from that initial estimate bore out the defendant's gloomy expectation. The initial estimate was more optimistic than the model's conclusion after it had simulated the consequences of the predicted interactions among the players. Remember that the changes from initial positions predicted in the game led also to a more accurate prediction about the expected final decision. In other words, the motivations and power of the plaintiffs and their cause suggested that my client would lose ground as negotiations wore on. The aura of doom and gloom was quick to return.

As figure 6.1 shows, my model predicted the negotiations would follow a complex path, first looking very encouraging and then turning sour. This figure illustrates what I meant earlier about this game being like multidimensional chess. If all that the players cared about was getting the result they advocated, this would be no harder than a round-robin chess tournament. But egos enter into negotiations and so enter into the negotiating game. Some players will take big risks to try to win big. Others are more concerned about not losing than they are about winning. That means figuring out which players are choosing their moves to get the plea they favor, which are picking moves that will get them the most credit for find-

FIG. 6.1. The Unengineered Plea Path

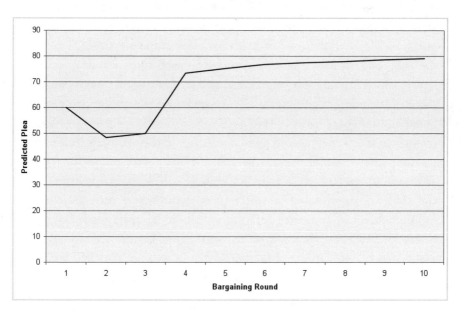

ing a settlement (or blocking one), and which are ambivalent about these
competing desires. Chess isn't this messy. Imagine trying to win at chess
when the rules for winning change with each opponent, as they do in the
negotiating game!

According to the model, the give-and-take in settlement discussions
would persist through eight meetings between the defendant's representatives and the U.S. attorney or his/her representatives. At the end of the
eight exchanges of views and arguments, the game indicates that the cost
of continued negotiations was no longer going to be worth the small
changes in the expected agreement. That agreement was just about at 80
on the plea scale—that is, one count of the severe felony plus several
lesser felonies.

The result anticipated by most of my client's lawyers and expected by
the board, and the plea predicted by the end of the game, were the same.
If my model couldn't improve on this, then my consulting partner and I
would have nothing to offer the firm that hired us. We would be just another expense.

However, the game uncovered something not anticipated by most of
the attorneys (one senior outside attorney had the ultimate result dead
right from the outset, although he/she did not quite see why it would arise
or how to get there). The model's simulation indicated that early discussions with the U.S. Attorney's Office and others would suggest that a powerful coalition of interests was going to form around a lesser set of
charges—almost like a gambit in chess, designed to suck the client and
the U.S. attorney into a move that would be used against them later. The
figure, which displays the predicted upshot of discussions at each stage in
the negotiations, anticipates that the second and third presentation of arguments by the defendant's side were likely to soften the U.S. attorney
enough so that he/she would contemplate agreeing to 50, a better mix of
multiple misdemeanor and felony counts not including a severe felony.
This was, in fact, the position believed to be held by the U.S. attorney at
the outset of negotiations, according to our interviews with the client's
representatives and attorneys. However, the initial analysis also showed
that with subsequent rounds of discussion, the U.S. attorney could be
and would be persuaded to take a tougher, not a softer, stance.

Why the move to a tougher stance after showing openness to a more
moderate settlement? The model indicated that the hard-line attorneys in

ABC and in the office of the U.S. attorney were going to execute a gambit. The ABC attorneys and the Department of Justice lawyers apparently were out to make names for themselves as tough guys who could bring evil corporations to their knees. The analysis suggested that they thought this was the perfect case to do it. Their gambit aimed to set up the U.S. attorney so that he/she would reveal softness early in settlement discussions. Then later they would pounce on this softness in order to embarrass the U.S. attorney politically, compelling him/her to take a tougher stance when it counted, in the end, lest he/she be tarred publicly with the ignominious tag of being soft on business malfeasance. That would not have played out well in the affected community or in the Department of Justice. To be sure, these lawyers were really rolling the dice, taking a big risk if their gambit failed. But they had every reason to think it would succeed—and probably it would have, had my client not been using a tool like the negotiating game.

Even though (according to the model) the U.S. attorney was willing to go for a mix of misdemeanors and lesser felonies, the logic showed that the hard-liner gambit would work. He/she would abandon a moderate stance, choosing instead to go for severe felonies. This was the U.S. attorney's way of solving the tough choice between pursuing the outcome he/she thought was right and following the gambit's path, thereby avoiding careerist costs and keeping the support of others in the Department of Justice and in the affected community. Seeing that the U.S. attorney had significant political exposure and careerist ego involvement in the case, it became apparent that we could find ways to gratify his/her ego and bollix the hard-liners' gambit.

The simulations uncovered several interesting patterns that opened the opportunity to engineer the outcome. First, as noted, the U.S. attorney took a tough stance according to the simulations because of pressure from attorneys at ABC and within the Department of Justice and in response to arguments from several stakeholders in the affected community. They were strongly committed to their point of view, and the U.S. attorney cared about being perceived by them as interested in helping them get justice as they saw it. This was particularly interesting because the U.S. attorney's own view of what could be supported as punishment through the judicial process was considerably less. The U.S. attorney's view involved none of the severe felonies and showed openness to at least some misdemeanors. The U.S. attorney, facing hard-line pressure, was willing to trade off some sense of legal correctness for political correctness and its attendant personal credit.

To engineer a better result, I simulated what might happen if the defendants altered their bargaining position from that of being prepared to accept numerous misdemeanors and maybe one lesser felony, a posture predicted to end in their caving in to one or more severe felonies. I looked at what would happen if they offered more concessions up front and also if they offered less. I also looked at how they could maneuver to get some important hard-liners to make arguments that would make those hard-liners look foolishly extreme to the U.S. attorney, turning the gambit on its head. In examining alternative strategies I took advantage of another insight gained from the base analysis: the U.S. attorney tilted more toward eagerness to make a deal than to sticking to a particular position. Also, it was evident that the defendant's strategy had to create leverage against the pressure exerted by the hard-liners who wanted a plea involving severe felonies.

It turned out that the best strategy for the defendant involved two shifts from the approach they had planned as reflected in the data they gave me about themselves. First, the one outside counsel who favored pleading guilty to one lesser felony and numerous misdemeanors needed to convey unity with the rest of the defendant's team in endorsing a plea to misdemeanors *only*. Although this one attorney had the right settlement in mind—that was the ultimate agreement on this aspect of the case— he/she could not so much as hint at this flexibility during the initial meetings with the U.S. attorney, and he/she did not.

This attorney acted out the scripted part perfectly. Jack Nicholson, great as his performance was in *A Few Good Men,* paled in comparison to the performance of the attorney who had to fake a commitment to a position he/she did not really believe in. Since he/she led negotiations on this issue, his/her ability to be convincing was critical, and convincing he/she was.

Of course, getting an attorney or anyone else to act contrary to their beliefs is no small order. It takes great faith that the model's logic should be allowed to trump personal intuition. As an old client used to say when introducing me to his colleagues, "Check your intuition at the door." The greatest value of a model is when it provides an insight that is contrary to the decision makers' expectations—when it correctly urges them to check their intuition at the door. It takes a courageous person to defy one's own beliefs and follow the lead derived from a computer model, since after all we never know what is or is not correct until after the fact. All we know is the model's track record for accuracy (but then everyone thinks his problem is unique) and whether the logic for the proposed action is persuasive. Fortunately, in this case the outside attorney being asked to change his/her approach had worked with my firm before on other cases. In fact, this attorney is the very person who persuaded the client to use my company's services. He/she had seen the model, as he/she put it, "work its magic" before, so this attorney had no problem agreeing to act out the part as written by the model.

The second maneuver that was required was considerably more challenging to "sell" to the defendant. The company's directors were naturally very concerned about this matter and were eager to find a solution. The simulations—remember, all of this analysis is happening before discussions with the U.S. attorney have begun—showed that the directors were so fearful of how the case was likely to unfold that they would cave in to the mounting pressure from the opposition hard-liners by agreeing to numerous felony counts including one count of the severe felony—that is, the hard-liners' chess gambit was going to work. To produce no severe felonies, getting instead multiple misdemeanors and one lesser felony as the plea, it was necessary to control the reaction of the board when the U.S. attorney pressed hard for an outcome the board was prepared to live with. The strategy for them was simple to articulate but hard to do: they had to take the position that they would not negotiate or authorize any discussion of felonies at all, risking the ire of the U.S. attorney and a breakdown in discussions.

They would have to repeat this message convincingly through months of negotiations between the U.S. attorney and their attorneys. Their attorneys would have to keep going to meetings with the U.S. attorney in which they repeated the message that they were unable to persuade the board to show some flexibility, followed by a plea that the U.S. attorney must give them the ammunition to convince their board that the matter could be settled. That ammunition was for the U.S. attorney to stand up to the hard-liner government attorneys, shooting down their arguments in front of the defendant's attorneys when they all sat in meetings together. That was the way for the U.S. attorney to flip the hard-liners' gambit on its head. To achieve this end, the board had to go along with the idea of their lawyers insisting on going to meetings with the hard-liners in attendance rather than trying to have private meetings with the U.S. attorney as they actually preferred.

Just imagine the board's reaction. Their first thought was "Who is this guy with the chutzpah [not a word they would have used] to tell us what to do when our entire business is on the line?" Indeed, when I proposed the optimal monetary offer to settle the case, an offer way below what they thought should be put on the table, they thought I was nuts. When I suggested that their attorneys should meet with the hard-liners as well as the U.S. attorney, they thought I was beyond nuts.

The model found that the optimal offer and settlement price was about a third of what the board thought they should make as their opening offer, and it found that this offer was best made in the presence of both the U.S. attorney and the hard-liners. The board members were sure the U.S. attorney would just get up and walk out if they did what the model was recommending. Fortunately, some of the other people involved in the case had worked with me, my partner, and the model before. They thought that the board should listen, and that they would listen if they just had an opportunity to confront the arguments head-on. With that in mind, the lead in-house attorney set out to arrange a meeting for me with a very senior executive from the firm. That senior representative would put the board's case before me and see if I could make convincing arguments based on the model's results—that is, could I persuade this executive that the strategy developed through the game was not nuts?

Mind you, I have faced this sort of situation many times. Anyone who works on behalf of my firm on a project is instructed never—let me repeat that word, *never*—to argue for or against an approach to a problem except based on results that can be pointed to directly in the model's output. We have no place for personal opinion. We are not experts on the substance of the problems we analyze, and generally we know little even about the industry involved, so there is no reason for anyone to take our personal opinions seriously. It is the very fact that we can show that the positions, tactics, and strategies we recommend come out of the model's logic and the client's data and not out of our heads that sells the client on the independence and the credibility of the view we express. They can argue with the model's logic—that's a conversation we love to have—and sometimes they do, but they always know that they will have to disagree with the logic or with the data inputs derived from our expert interviews (usually with their own experts), and not with us. Logic and evidence, not anyone's personal opinion, are the focus of our presentations, briefings, and discussions and are, in the end, the basis on which people should decide whether to try an approach contrary to their initial intuition.

No one should blindly follow a model. It is, after all, just a bundle of equations. But neither should people dismiss a model's results out of hand just because its implications and their personal opinions differ. To repeat myself, the model's greatest value is that it provides clients with a different way of thinking about their problems. That is an important part of the power that game theory, strategic thinking, brings to the table. Fortunately, as it turns out, clients generally find that the logic makes sense to them, and the data, after all, are theirs and can be adjusted and rerun to see how robust the findings are. So in the end, when they listen to us it is because they are sold on the integrity of the process.

Of course, not all clients "check their intuition at the door." When they don't follow the model's advice, the model's prediction for what is likely to follow in that case tends also to be accurate. They end up with an outcome that is typically a lot less favorable. In the case at hand, the board's senior management representative was sold on the model-based advice after examining me and the results for about eight grueling hours. The board agreed to go ahead with the approach we recommended.

To communicate the message credibly, the firm's general counsel agreed to meet with the U.S. attorney, and requested, even insisted, that the hard-liners be present. Naturally, this insistence came as a surprise to the U.S. attorney. It was an even bigger surprise and a deep disappointment to the hard-liners. According to the model, they were eager to be backstabbers behind the scenes, sheltered from the light of day. They wanted to take a shot at influencing, maybe even cajoling or coercing, the U.S. attorney after he/she met privately with the firm's general counsel. They tried to get out of attending a meeting with the general counsel, claiming scheduling conflicts. The firm's general counsel deflated that maneuver by stipulating that he/she would meet whenever it was convenient for them. Their hopes for executing their gambit were crushed. By our surprise move, they lost that opportunity.

The general counsel conveyed the board's (sincerely held) conviction that they had done nothing that warranted criminal charges beyond mis-demeanors. The general counsel went on to argue that the firm's agreeing to what the U.S. attorney was demanding was tantamount to giving up the affected, important part of their business. The message was that the board would not even contemplate any sort of deal that involved pleading guilty to felonies.

Of course, we knew that if push came to shove, the firm would have caved in and accepted one or more counts of the serious felony even if doing so would destroy an important part of their operations. They would have done so to bring the process to a speedy end so that they could get on with the rest of their business. They believed that even if they pre-vailed in court—recognizing that this is always uncertain even in the best of circumstances—the political, social, and economic costs of a pro-longed trial would be unbearable. It was better to settle on a plea agree-ment and suffer the consequences. Remember Arthur Andersen (not the firm involved in this case). The company fought the charges against it re-sulting from its audits of Enron and was found guilty in court, only to have that judgment overturned by the Supreme Court after it was essentially no longer in business. Sometimes it is worse to win in court than to accept a plea agreement even when you are innocent. The judicial process may eventually come to the right answer, but too often the right answer only comes after unendurable costs have been borne by the defendant.

The model's maneuvers, in this as in many other situations, are designed to prevent push from coming to shove. The approach recommended through the model's logic is indeed the very kind of bluffing we were talk-ing about in our exploration of fundamental game theory back in Game Theory 101 and 102. We knew the board would plead to more than mis-demeanors if left to its own devices, but the U.S. attorney did not and could not know that.

After months of discussion, as anticipated, the U.S. attorney responded to the pressure from the board of directors and chose sides in the ongoing war. Instead of capitulating to the hard-liners within the government, the U.S. attorney berated them in the meeting we had insisted they attend. The U.S. attorney seized the chance to reinforce his/her own initial view of what constituted a just agreement. The hard-liners were made to look mean-spirited and unrealistic, even foolish in their demands. The case was settled by the defendant's pleading guilty to several misdemeanors and one lesser felony. This was the outcome they desired and felt was just. They believed it was out of reach, and it would have been had they gone in asking for this agreement. Going in with the final result would have left them feeling compelled to make more concessions. As it was, they got an outcome vastly better than the defendant's management or attorneys (save one) thought possible.

This case typifies the engineering process. The board of directors, the U.S. attorney, the Department of Justice hard-liners—they are, obviously, all different people, with different upbringings, personal experiences, and beliefs about the world. But they all make choices across the same dimensions of human behavior—there are, after all, only so many options to choose from. They can look for compromises; they can try to coerce people into capitulating to them; they can surrender to their adversaries; they can lock themselves into a war against their opponents; and they can bluff doing some of the above. That's about the full array of choices of action for any problem. The key to this case was isolating the U.S. attorney; he/she was the principal driver of the outcome. What he/she thought was "right"—whether it was or was not in any absolute sense—and what he/she wanted in ego satisfaction: these were the two essentials behind engineering the settlement. The real question of this case was how those two interests could be aligned most favorably for my client.

If I were to have brought the area-specialist mentality to this engagement, then perhaps I would have read through the thousands of pages of background produced by the armies of lawyers, perhaps I would have delved into volumes of case law, and perhaps I would have ultimately produced a brilliant argument as to why my clients deserved only misdemeanors and a minor felony. (Of course, this is not what my client hired me for, and is in fact exactly what the lawyers would have produced if the matter had landed in court.) But then, in that clash over the merits, however brilliant an argument I could have produced, despite all of my efforts, it probably wouldn't have meant a darn thing. Because, of course, the talented government attorneys on the other side could have produced just as sparkling a case for severe felonies—and that leaves so very much to chance. No, the path to favorable resolution was in doing the work to produce an accurate understanding of the lay of the land, and then finding ways to work *with* it, not against it, through sequences of interaction.

The process of predictioneering does not rely on the recounting of grievances (which all too often only hardens positions). If people are congregated around an impasse, well, it's indeed unfortunate that there's a conflict, but the very act of congregating around it suggests that the parties are in search of some dynamic that will yield an outcome—whatever it may be—to break the stasis. Predictioneering provides a complex network of decision-making channels, valves that open and close as actions are taken or passed upon. If I offer options A, B, and C, then doors D, E, and F open, and so on exponentially (again, hence the need for the trusty computer!). Suddenly, as this network routes decisions, grounds shift, positions change, and in this case a U.S. attorney is led to a place where he/she feels both validated in his/her own views and accomplished in forcing a little heavier penalty on my client than it was apparently (but not actually) initially willing to endure.

This process involves exploiting or altering people's perceptions of a situation by looking within the model's round-by-round output to work out who is responsible for shifts in positions and how to counter those shifts if they have bad consequences for the client. The process is no different whether the problem is resolving Iran's nuclear program, figuring out what al-Qaeda is likely to do, or facilitating the merger of companies. Every one of these situations involves humans who are not all that different from one another, regardless of where they go to sleep at night.

So in the next chapter I'll look at a few current problems such as those listed above to see how we might engineer beneficial outcomes. The examples will help illustrate the potential costs of failing to see or to address what may be around the corner.

7

■

FAST-FORWARD THE PRESENT

ONE OF THE great benefits of being affiliated with Stanford's Hoover Institution is the opportunity to participate in small seminars with some of the world's most interesting scholars and policy makers. These seminars are often off the record, which means that there is the chance for frank exchanges of views on important issues of the day. The discussion during one such seminar, on the Israeli-Palestinian dispute, led me to consider how game-theory reasoning might contribute to tackling the seemingly insurmountable obstacles to peace. The approach I thought about is not a solution to the dispute, but it is a potentially useful step toward advancing the real prospect of a lasting peace.

For all of its limitations, the idea I came up with provides an example of how game-theory reasoning can nudge us in a new direction even under the most seemingly intractable circumstances. If game-theory logic can foster progress on the Israeli-Palestinian dispute, it surely will have contributed to solving one of the most important foreign policy problems of our time. With that in mind, let's have a fresh look at Israeli-Palestinian relations. And who knows, maybe somebody reading this book can help turn the idea into reality or can point out some fatal flaw in it.

LET'S MAKE A DEAL

Land for peace and peace for land are two formulas that are doomed to failure, whether in the Middle East or anywhere else. It's an idea that sounds sensible, so it attracts lots of attention. Ehud Barak proposed a land-for-peace deal at the July 2000 Camp David summit between him (he was Israel's prime minister), Yasser Arafat (then president of the Palestinian Authority), and President Bill Clinton. The Oslo Accords in 1993 also were a land-for-peace deal. Barak's Camp David plan and its later variants failed. The Roadmap for Peace, another land-for-peace arrangement, failed too. All land-for-peace or peace-for-land deals by themselves will do the same. They are no way to end violence, because neither assures either side that the other is making a lasting promise, a credible commitment.

Each promise—land for peace or peace for land—suffers from what in game theory is sometimes called a time inconsistency problem. That is, one party gives an irreversible benefit to the other party today in the hope that the other will reciprocate tomorrow. Almost certainly instead, the side getting the irreversible benefit exploits it to seek even more gains before delivering on its promises. Giving up land on the promise of peace inevitably leads to demands for more land before peace is granted. Giving peace on the promise of land later has much the same problem. The peace giver lays down its arms to show good faith, but then the land giver is free to renege, feeling no compulsion to follow through with land the opponent can no longer take.[1]

Time inconsistency problems arise in many contexts, not just land for peace or peace for land. In fact, we saw an example of this problem earlier when we discussed North Korea. The threat of reneging on promises once an adversary has disarmed is exactly why asking Kim Jong Il to dismantle his nuclear capability won't work, but negotiating a deal in which he agrees to disable his nuclear program can work. The problem is no less consequential in the Middle East.

Look at the decision by Israel's hard-liner prime minister Ariel Sharon to withdraw (unilaterally) from Gaza in August 2005, ceding that territory to the Palestinians. An important part of Sharon's motivation seems to be that he concluded it was too costly to defend Gaza. So he chose to make Israeli settlers abandon their homes, whose existence was a major flashpoint with Palestinians, in the hope that yielding Gaza would help promote goodwill and peace. The belief that good deeds, whatever their motivation, will elicit a good response reflects optimism about human nature that sometimes is met by reality but all too often is met instead with greed and aggression. As you know, game theory rarely takes an optimistic view of human nature. Sharon's optimism was predictably wrong. Shortly after the democratically elected government run by Hamas in the Palestinian Authority used force to oust Palestinian president Mahmoud Abbas's Fatah from Gaza, Hamas increased missile attacks against Israeli towns near the Gaza border. Land, freely given with no strings attached, did not produce peace. It produced a demand for more land and an increase in violence.

Mind you, this failure on behalf of the pursuit of peace is not some particular flaw among Palestinians. The Israelis have done much the same in the past. Having defeated their Arab rivals in the 1967 war and then again in 1973, Israel not only occupied previously controlled Egyptian, Syrian, Jordanian, and Palestinian land but also allowed the spread of settlements presumptively justified by the biblical covenant between Abraham and God. In fact, Israeli settlements almost always occupy the high ground surrounding Palestinian villages, making it all but impossible for Palestinians to enjoy a sense of security within their own homes. And even more troubling, Israel for decades restricted the movement of Palestinians into and out of Israel, just as they had done within Israel to Israeli citizens of Palestinian extraction. The upshot is that the Israeli government prevented Palestinians from following a peaceful road to independence by restricting their freedom of assembly. When Israel had the opportunity to promote peace with Palestinians after the 1967 war, it fell short, just as the Palestinians have fallen short in efforts to promote peace with the Israelis.

Every land-for-peace and peace-for-land formula I know of has ended in failure. Every such effort, whether unilateral, bilateral, or multilateral, has, if anything, made the situation worse by raising false hopes only to see them dashed. They are always dashed because the peacemakers simply do not pay attention to the time inconsistency in their strategy. They rely on goodwill and building trust when there is neither. Instead, they should leverage progress toward peace on the narrow self-interest of the contending sides. They should consider whether what they propose is a self-enforcing strategy from which no one has an incentive to deviate, rather than looking for an approach that requires mutual cooperation. Remember the prisoner's dilemma from Chapter 3? Both players in that game are better off if they coordinate with each other to pursue joint cooperation instead of ending up competing with one another. The problem is that each is even better off by not cooperating if the other player chooses to cooperate. The upshot is that neither cooperates, leaving both worse off than they could have been. That, in fact, is the dilemma. Like the Israeli-Palestinian dispute, joint cooperation is not a sustainable solution unless the structure of the game changes first.

One way to change the game is to make costs and benefits change directly and automatically in response to the actions chosen by each player. A self-enforcing strategy solves this problem and can help promote peace and prosperity for each side. Here I would like to use the power of game-theory thinking to propose an important step toward peace between Palestinians and Israelis. It is not a comprehensive peace plan, but it is a way to make peace more likely. What I will say follows logically from game-theory reasoning, but it is not a mere assessment of what is likely to happen. It is a statement of logic in support of a way to end violence. It is a prescription for progress.

Key to my proposal is the phrase "self-enforcing." It is an arrangement that requires little or no cooperation or trust between Israelis and Palestinians. The idea I have in mind provides each side with incentives to promote peace and resist terrorism purely in their own interest and utterly without concern for whether it helps the other side. In that sense it follows game theory's dismal view of human nature.

My idea is that the Israeli and Palestinian governments will distribute a portion of their tax revenue generated from tourism (and only from tourism) to each other. Before going into the details—where the devil resides—let's first see why tourist revenue and not any other. Why not, for example, promote peace by setting up joint Israeli-Palestinian ventures, or allowing freer movement between regions, or some other scheme? As we will see, shared tourist revenue provides a nearly unique opportunity.

The Palestinian Authority (PA) leadership routinely identifies tourism as one of the major pillars of the future Palestine's economy. This is an eminently reasonable expectation given the vast number of historic and religiously important sites within the current and expected future territory of Palestine. The PA's gross domestic product in 2007 was $4.8 billion. During peaceful periods, tourism represents more than 10 percent of income, and it could be much, much higher. By comparison, Israel's GDP in 2007 was over $160 billion. Its tourist revenue was $2.9 billion in 2005 and $2.8

billion in 2006, and it is expected to be around $4.2 billion in 2008. So tourist revenue is a nice but relatively modest source of Israel's income.

Tourism has a feature that can be exploited to improve the prospects of peace. You see, tourism and the tax revenue generated from it are highly sensitive to violence. For example, take a look at figure 7.1. The horizontal axis shows the range of violent Palestinian and Israeli deaths resulting from their conflict for the years from 1988 until 2002.[2] The scale reflects a range of quarterly deaths from 0 to about 300. The vertical axis shows the number of tourists (in thousands) who visited Israel each quarter between the same period, 1988 to 2002.[3] Unfortunately, I have not located comparable tourist data for the PA, but I have found enough to see that the pattern is the same. When violence goes up, tourism goes down, and when violence drops, tourism returns.

The line in the figure shows the estimated rate of tourist response three months after the reported level of violence, while the dots show the observed amount of tourism in Israel associated with actual violence that

FIG. 7.1. How Much Does Israeli Tourism Respond to Violence?

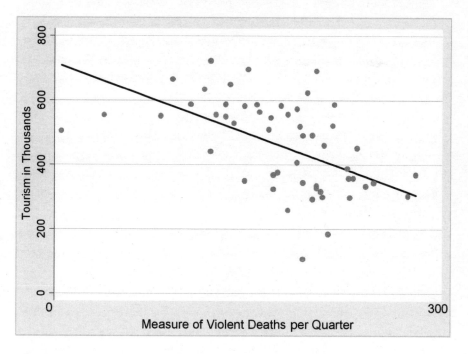

occurred three months earlier. Tourism is delayed by three months after observed violence to give prospective visitors enough time to change their plans.

Clearly, more violence means a lot fewer tourists. In fact, on average, every violent death translates into 1,300 fewer tourists and 2,550 fewer hotel bed-nights sold to tourists. There were 53 violent deaths during the typical quarter covered here. That translates into nearly 70,000 fewer tourists in a three-month period suffering average violence compared to the number expected in quarters with no dispute-related violent deaths. Israel averages about 450,000 tourists per quarter, so 70,000 is a serious number. With Israel enjoying about $3 billion in tourist revenue in an average recent year, and an average year having about 280,000 fewer tourists than might conservatively be expected in a peaceful year, that translates into about $500 million tourist dollars lost each year, not counting whatever is also lost by hotels, restaurants, taxis, car rental companies, guides, and so forth on the Palestinian side of the border.

And what is the Palestinian experience like? As I mentioned, it is harder to get equivalent data for the PA, but still there is plenty of evidence that the picture is grim. For instance, there were about 90 hotels in the PA before the intifada that started in late 2000. By the end of 2001 the number had dropped precipitously to around 75. Naturally, hotel openings and closings result from how much business they do. Figure 7.2 shows the number of hotel guests in the PA for each year from 1999 through 2005, as reported by the Palestinian Authority.[4] The intifada produced a quick, intense, and easily anticipated response: hotel stays—and tourism in general—plummeted. Estimates from the PA suggest a loss of 600 million tourist dollars between the second intifada's inception in September 2000 and July 2002. Annual PA tourism revenue in that same period was only $300 million, so the loss was as large as the total tourist revenue. Keep these numbers in mind. We will return to them.

With these facts under our belts, we can work through the game-theory logic that points to the attractiveness of tourist tax dollars as a path toward peace. Imagine, for instance, that President Obama's government or the United Nations presses the Israeli and Palestinian Authority governments to share with each other tax revenues arising exclusively from tourism and then administers the distribution of the funds. Each side's share of those

FIG. 7.2. Tourism in the Palestinian Authority Since the 2000 Intifada

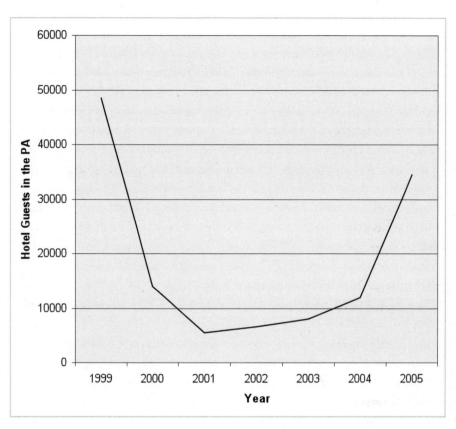

tax dollars is to be in direct proportion to its current proportion of the total Israeli and Palestinian populations in the PA and Israel.

The tourist tax revenue arrangement need not last forever. It must include an irrevocable commitment for it to persist for a long time (say twenty years), and it is important that this tax revenue sharing be tied to a fixed formula based on the current populations and not on future changes in those populations. Opening the formula to renegotiation could create perverse incentives. It is also essential that the definition of tax revenue originating from spending by tourists be based on predetermined rules for estimating this source of income. Independent accounting firms might be used to provide a standard way to define and identify tourist revenue and the taxes derived from it. This tax revenue would then be allocated to each side over the agreed duration of the program based on a onetime fixed population-based formula with no questions asked.

Some tourist-derived tax revenues are obvious. Hotels check the passports of foreign visitors, and so, as we saw, it is straightforward to know how many tourists checked in, what their hotel bills were, and what were the tax portions of those bills. For every foreigner staying in a hotel—whether strictly a tourist or claiming some business purpose—one might stipulate that the taxes on that hotel stay go into the tourist tax revenue pot. It will be necessary to devise a good monitoring system to prevent underreporting, but that is probably something governments already have dealt with.

Restaurants don't have as obvious a way of determining who their foreign guests are, but perhaps accountants can find a clever way to approximate the percentage of a restaurant's taxes that is attributable to tourists. This might depend on the location of the restaurant, the proportion of foreign hotel guests in the area, the location of the bank for payments made by credit card, or many other criteria. The same might be true for shops. Those selling souvenirs, for instance, are likely to have more tourist-based revenue than those selling groceries. At passport control, visitors declare whether the purpose of their visit is business or pleasure. There, too, it is possible to create a revenue formula that approximates how much is spent by those saying they are tourists. Anyway, I am not trying to do the job of accountants, and I certainly am not qualified to do so. Accountants will be good at setting up sensible rules to identify the relevant sources of tax revenue, especially if their fees are also tied to that revenue.

On the population side, if Palestinians currently make up 40 percent of the population in the area, then 40 percent of the tourist tax dollars (or shekels, or any other currency) automatically goes to the Palestinian Authority's recognized government and 60 percent to the recognized Israeli government. Government recognition could be determined by who selects the personnel in the Palestinian and Israeli delegations to the United Nations. This avoids the risk of a dispute over who constitutes the relevant government, since general diplomatic recognition is itself a contentious issue between the Israelis and Palestinians. Just who the Palestinians and Israelis "are" should be defined in terms of people residing in the area and should not include diaspora populations. To include them will result in the meaning of "Palestinian" and "Israeli" being manipulated for political and economic gain. The money distributed is then dictated by how many tourists come and how much they spend, regardless of whether they spend more or less in Israel, Palestine, or disputed territory. Furthermore, there are no restrictions on how the Israelis or Palestinians spend the money received under this program. If leaders invest this money in improving the quality of life for their people, that's great. If they want to sock it away in a secret bank account, that is between them and their constituents. The key to success is that the money is neither given as a reward for advancing peace nor withheld as a punishment for hindering peace.

Recall the current situation. Israel expects to bring in about $4 billion in 2008 from tourism and expects, with peace, that this will grow sharply over the next several years. Based on the evidence in figure 7.1, we can estimate that tourist revenue would grow 50 percent relative to what it has been since 2001 if peace prevails. Probably Palestinians in the PA would experience an even greater increase in tourist earnings, as tourism seems to respond more sharply to violence in their region than in Israel. This is not surprising, given that they bear the brunt of the deaths. So, with peace, Israeli revenue would rise from about $4.2 billion in 2008 to (at least) $6.3 billion once a lasting peace was established. Palestinian revenue from tourism would probably increase from $300 million to (at least) $600 million, the amount they earned from tourism in 1999, the year before the second intifada began.

Current total tourist revenue with ongoing violence is reasonably estimated at $4.5 billion for Israel and the PA combined. With peace, the estimate rises to $6.9 billion. Assuming a 20 percent average tax rate on tourism earnings, this means a tax revenue pot worth $1.4 billion with peace compared to $900 million with ongoing violence.

Without a tax-sharing agreement and without peace, Israel's tax revenue (continuing to assume a 20 percent tax take) from tourism is projected to be $840 million in 2008 or 2009, while the PA's tax revenue from tourism is projected to be $60 million. With a 60-40 revenue sharing arrangement and peace, Israel's tax take from tourism is projected to be at least $830 million—essentially a wash in terms of tax earnings between war and peace. The PA's projected tax take under the proposed plan is $550 million, a more than ninefold increase, not to mention an increase in total revenue for the PA of around $1 billion, equivalent to a 20 percent increase in GDP. That's serious economic growth!

Okay, so we've seen the numbers. Now let's follow the logical stream.

What we have seen is that so long as terrorist attacks or other forms of violence persist at a significant level, far fewer tourists visit Israeli-controlled or Palestinian-controlled sites. When there is significant violence, there will be little tourist income to distribute. Thus, if the Palestinian leadership does not engage in effective counterterrorism, it will get few, if any, funds. It will not be deprived of funds, as it is so often, merely because the Israelis or the international community do not like its policies. Money will not be withheld or payment made contingent on the Palestinians doing what anyone else demands. Money will flow or dry up purely because tourists abhor violence in the places they want to visit. If the Palestinians crack down on the sources of terrorism or other forms of attacks against Israel, then the decreased violence will almost surely be followed by a significant increase in tourism. If tourism increases ever so slightly more than my conservative estimates, decreased violence will mean more tax revenue for both governments.

Even with my conservative calculations, PA revenue skyrockets and the Israelis lose nothing. Similar gains can be realized if the Israelis control actions by settlers and other groups that may be inclined to foment trouble with Palestinians, leading inevitably to retaliatory strikes back and forth. The governing parties on each side should have the right incentives to prevent that result.

A failure to engage in effective counterterror or proper policing leaves the status quo in place. Successful counterterrorist policies and effective policing enrich both sides without either side's having to cooperate directly with the other. Of course, it is likely that collaborative efforts between the intelligence services of each side would emerge to enhance the income of both. There is no mechanism in this proposal by which one side can improve these revenues without also improving the revenues flowing to the other side. That is why the plan is self-enforcing and potentially equally beneficial to both sides.

Obviously, the Israeli-Palestinian conflict is not only about their respective economies. And equally obviously, many economic incentive approaches have been proposed before, going back at least to the time of Winston Churchill. However, earlier economic schemes assumed investment strategies that required mutual coordination and cooperation, and gave the investing side—the Israelis—the ability to pull the plug if they didn't get what they wanted. Such approaches fail to assign equal responsibility to the Palestinians and the Israelis for enforcing peace or for punishing violators of the peace. Such approaches also foster a fear of economic dependence in Palestinians employed by Israeli-funded businesses. Sharing tourist tax revenues has none of these limitations. Incentives are symmetric, and the responsibility for enforcement is also symmetric. Violations of the peace by either side mean a loss of revenue for both sides.

Not only is tourism important for the future PA economy, but the Palestinian Authority's leaders have also shown that they can and will control threats to the peace that directly impact some forms of income. For example, in the case of the gambling casino in Jericho controlled by the Palestinians, we know that they have successfully secured the road to the casino. That road to casino riches experienced minimal security threats even at the height of the intifada. By securing the road, the Palestinians also ensured the flow of revenues from it. Additionally, we know that the least developed tourist areas in Jerusalem are in the Palestinian quarters, with a similar pattern of underdevelopment of tourist facilities throughout the Palestinian Authority's territory and in areas in Israel dominated by Palestinian residents. With peace we can expect that international hotel chains, restaurants, boutiques, and other enterprises catering to the tourist trade will grow disproportionately in the underdeveloped areas, thereby making the tie between tourism-generated prosperity and peace all but self-funded in the mid- to long term.

Finally, sharing tourist tax revenue (recalling that what is shared will by its nature be minimal if there is no peace) will promote a "confidence building" step that requires no trust on either side. It also should promote more counterterrorism efforts from the Palestinian side and probably fewer new settlers on the Israeli side. If this revenue-sharing scheme helps pacify the area and helps promote effective counterterrorism, then the door is opened for more encompassing negotiations over fundamental issues. Terrorist movements, once destroyed, rarely reemerge. The revenue-sharing concept can be a way to move the "peace process" in a positive direction without relying on mutual trust, or even mutual contact, for the time being.

If religion truly dominates the divide between Israelis and Palestinians, as many believe, then the tourist tax plan will fail to promote peace, but it will reveal who the main impediments to peace truly are. Identifying who exactly is willing to sacrifice their own people's economic and social well-being for religious or other reasons will make it easier to know with whom to negotiate and with whom it is a waste of time. That, in turn, can open the way for fragmenting resistance to peace and make subsequent counterterror efforts more focused and more effective. Either way, such a self-enforcing quest for peace is unlikely to make the situation worse, and has a good chance of making it better.

Some people resist ideas such as my tourist-tax plan because they are sure they can't work. They think that the cultural divide between Israelis and Palestinians or Jews and Muslims is just too deep to respond to ordinary economic incentives. They think there is a culture of violence that cannot be overcome. They think this even as they see that Northern Ireland is no longer racked by daily explosions and that tens of thousands of former insurgents in Iraq are sticking to their new role as Concerned Local Citizens. Maybe they are right in the case of Palestinians and Israelis, but history does not support their assumption.

Throughout the long history of Muslim domination of the Middle East until roughly the start of the Second World War, Jews lived better and more freely among Muslims than almost anyplace else in the world. When the so-called Moors controlled Spain, Jews enjoyed the tolerance of the Muslim leadership. That tolerance came crashing down when Ferdinand and Isabella unified Spain under Catholic rule. The basis of Palestinian-Israeli conflict resides, at least for many, in economics, not religion. Religion is a politically useful and easy organizing principle that unscrupulous people use to marshal support, but it is not what the fight was or is primarily about. The fight is about land in a locale where, for most, the economy was historically tied to owning property, just as it is in all traditional societies. The economies in the territories lived in and controlled by Israeli and Palestinians still rely significantly on land, but not nearly as much as they did decades ago. Israel has a modern economy in which agriculture plays a much altered role. The Palestinians aspire to a significant degree to have a modern, service-based economy in tourism. These are the conditions that are ripe for a self-enforcing incentive plan.

Naysayers are too quick to equate what they see people do with what they think their core values are. Because terrorist acts seem so extreme, so fanatical, so incomprehensible, many of us are quick to assume that terrorists are a breed apart. They are thought of as people who cannot and will not respond to rational arguments. And yet we already know that even al-Qaeda insurgents in Iraq can be induced to change their ways for just ten dollars a day. Put the history of Jewish-Muslim relations together with the responsiveness of former insurgents to modest economic rewards, and it's hard to see the downside to trying a new economic approach, especially one that promises virtually no economic downside for one party and huge gains for the other. As the old anti-war song says, "Give peace a chance."

Even those who absolutely cannot believe that Palestinians or Israelis would value economic incentives over religious principles should want this tourism-incentive plan tried out. Why? Because it has a "hidden hand" benefit alluded to earlier that directly addresses the concern of naysayers. I think we can all agree that there are some hard-liners on the Palestinian side who don't care about building a strong Palestinian economy, and others on the Israeli side who are certain God did not intend the land to be occupied by anyone other than Jews. These hard-liners will do whatever they can to thwart peace. They will foment violence to prevent tourists from coming. But we should also be able to agree that there are at least some pragmatists on each side as well. The revenue-sharing strategy will ensure that the pragmatists have a strong incentive to identify hard-liners and fight them. The pragmatists will have an incentive that they do not currently have to provide counterterrorism intelligence to their governments in order to ferret out the hard-liners and stop them from interfering with the massive economic improvements promised by this plan. Thus, it should become easier to find and punish the hard-liners, thereby strengthening the hand of pragmatists on both sides. That's something that should appeal to those who fear the power of the hard-liners.

What I want more than anything to show in this book, and hope that I have done so to a degree so far, is that by thinking hard about the interests involved in a given problem, we have the opportunity to take the *best available* steps to ensure optimal outcomes. As this next example will show, when we are unaware of the interests at play, or willfully ignore them, we can invite ruin upon ourselves.

INCENTIVIZING IGNORANCE

Arthur Andersen was driven out of business by an aggressive Justice Department looking for a big fish to fry for Enron's bankruptcy. Later, on appeal, the Supreme Court unanimously threw out Andersen's conviction, but it was too late to save the business. Thousands of innocent people lost their jobs, their pensions, and the pride they had in working for a successful, philanthropic, and innovative company. Andersen's senior management apparently was entirely innocent of real wrongdoing. Unfortunately, they nevertheless helped foster their own demise by not erecting a good monitoring system to protect their business from the misbehavior of their audit clients. In fact, that was and is a problem with every major accounting firm. In Andersen's case, I know from painful personal experience how needless their sad end was.

Around the year 2000, the head of Andersen's risk management group asked me if I could develop a game-theory model that would help them anticipate the risk that some of their audit clients might commit fraud (this is where my work related to the Sarbanes-Oxley discussion from a few chapters back began). As I have related, three colleagues and I constructed a model to predict the chances that a company would falsely report its performance to shareholders and the SEC. Our game-theory approach, coupled with publicly available data, makes it possible to predict the likelihood of fraud two years in advance of its commission. We worked out a way to identify a detailed forensic accounting that helps assess the likely cause of fraud—if any—as a function of any publicly traded company's governance structure.

We grouped companies according to the degree to which our model projected that they were at risk of committing fraud. Of all the firms we examined, 98 percent were predicted to have a near-zero risk of committing fraud. Barely 1 percent of those firms were subsequently alleged to have reported their performance fraudulently. At the other end of our scale, about 1.5 percent of companies were placed in the highest risk category based on the corporate organizational and compensation factors assessed by the model. A whopping 85 percent of that small group of companies were accused by the SEC of committing fraud within the time window investigated by the model. This is a very effective system that produces few false positives—alleging that a company would commit fraud when it apparently did not—and very few false negatives—suggesting that a company would not commit fraud when it subsequently did.

Enron was one of the 1.5 percent of companies that we highlighted as being in the highest risk category. You can see this in the table on page 119, which shows our predictions for a select group of companies that eventually were accused of very big frauds. The table shows our assessment of the risk of fraud for each company each year. The estimates of interest are for 1997–99. These assessments are based on what is called in statistics an out-of-sample test. Let me explain what that means and how it is constructed.

Suppose you want to know how likely it is that a company is in either of two categories: honest or fraudulent. Using game-theory reasoning, you might identify several factors that nudge executives to resort to fraud when their company is in trouble. A few chapters back we talked about some of those factors, such as the size of the group of people whose support executives need to keep their jobs, and we talked about factors that provide early-warning signs of fraud, such as dividend and management compensation packages that are below expectations given the reported performance of the firm and its governance structure.

We know that some conditions, including the amount paid out in dividends, indicate whether fraud is more or less likely; but how important is the magnitude of dividend payments in influencing the risk of fraud compared to, for instance, the percentage of the company owned by large institutional investors? That too is an important indicator of the incentive to hide or reveal poor corporate performance. There are statistical procedures that evaluate the information on many variables (the factors identified in the fraud game devised by my colleagues and me, for example) to work out how well those factors predict the odds that a company is honest or fraudulent (or whatever else it is that is being studied).

There is a family of statistical methods known as maximum likelihood estimation for doing this. We won't worry here about exactly how these methods work. (For the aficionados, we used logit analysis.) The important thing is that these methods produce unbiased estimates of the relative weight or importance of each of the factors, each of the variables, thought to influence the outcome. By multiplying each factor's value (the number of directors, for example, or the percentage of the firm owned by institutional investors) by its weight, we can get a composite estimate of the probability that the firm will be honest or will commit fraud two years in the future. If the theory is just plain wrong, then these statistical methods will show that the factors in the equation do not significantly influence whether a firm is honest or fraudulent in the way the theory predicts.

The weights we estimated were derived from data on hundreds of companies for the years 1989 through 1996. Since the thing we were interested in predicting—corporate honesty or fraud—was unknown for 1997 in 1995, or for 1998 in 1996, or for 1999 in 1997, and so forth, the statistically estimated weights were limited to just those years for which we knew the outcomes as well as the inputs from two years earlier. Thus, the last year for which we used the statistical method to fit data to a known outcome was 1996. We then applied the weights created by the in-sample test to estimate the likelihood of fraud for the years of data that were not included in our statistical calculation. Those years are the out-of-sample cases. The out-of-sample predictions, then, cover the years 1997 forward in the table. Of course, since this analysis was actually being done in 2000 and 2001, we were "predicting" the past.

Now, you may well think this is an odd view of prediction. It is unlike anything I have discussed so far. What, you might wonder, does it mean to *predict* the past? That must be very easy when you already know what happened. But remember, in the out-of-sample test, nothing that happened after 1996 was utilized to create variable weights or to pick the variables that were important. Since the predictions about events after 1996 took no advantage of any information after that year, they are true predictions even though they were created in 2000 and 2001. This sort of out-of-sample test is useful to assess whether our model worked effectively at distinguishing between companies facing a high and a low risk of fraud. That is not to say that it is useful from a practical standpoint, even though it is useful from the perspective of validating the model and in providing confidence about how it could be expected to perform in the future. Let me explain what I mean:

Predicting the past can be helpful in terms of advancing science, even though it is not of much practical use when it comes to avoiding the audit of firms that have already committed fraud. But if it accurately predicted the pattern of fraud in the past, it is likely to do the same in the future.

Here's another way to think about this: The fraud model uses publicly available data. If Arthur Andersen had asked my colleagues and me to develop a theory of fraud in 1996 instead of in 2000, we could have constructed exactly the same model. We could have used exactly the same data from 1989–96 to predict the risk of fraud in different companies for 1997 forward. Those predictions would have been identical to the ones we made in our out-of-sample tests in 2000. The only difference would have been that they could have been useful, because they would then have been about the future.

SAMPLE OF BIG FRAUD PREDICTIONS

Company	'91	'92	'93	'94	'95	'96	'97	'98	'99
Bank of America	ND	ND	1	5	2	5	5	1 F	1
Boston Scientific	ND	ND	4	4	4	4	5 F	NDF	ND
Cendant	ND	1	5	4	ND	4 F	1 F	1	1
Cisco	ND	ND	ND	4	ND	4	ND	4	NDF
Enron	2	1	4	2	3	4	5	5 F	5 F
Informix	1	4	4	4 F	5 F	5 F	3 F	ND	5
Medaphis	2	2	3	4 F	5 F	5 F	4 F	2	ND
Rite Aid	2	2	1	3	4	5	4	5 F	3 F
Waste Management	5	5 F	5 F	4 F	3 F	5 F	5 F	2	1
Xerox	1	1	1	1	3	3	5	5 F	5 F

Table notes: Prediction: 1 = very low risk, 2 = low risk, 3 = moderate risk, 4 = high risk, 5 = very high risk; ND indicates we did not have sufficient data to make an estimate; F shows the years for which fraud was subsequently alleged to have occurred, a piece of information unknown at the time of our predictions.

Clearly, we had a good monitoring system. Our game-theory logic allowed us to predict when firms were likely to be on good behavior and when they were not. It even sorted out correctly the years that an individual firm was at high or low risk. For instance, our approach showed "in advance" (that is, based on the out-of-sample test) when it was likely that Rite Aid was telling the truth in its annual reports and when it was not. The same could be said for Xerox, Waste Management, Enron, and also many others not shown here. We could identify companies that Andersen was auditing that involved high risks, and we could identify companies that Andersen was not auditing that they should have pursued aggressively for future business because those firms were a very low risk. That, in fact, was the idea behind the pilot study Arthur Andersen contracted for. They could use the information we uncovered to maintain up-to-date data on firms. Then the model could predict future risks, and Andersen could tailor their audits accordingly.

Did Andersen make good use of this information? Sadly, they did not. After consulting with their attorneys and their engagement partners—the people who signed up audit clients and oversaw the audits—they concluded that it was prudent *not to know* how risky different companies were, and so they did not use the model. Instead, they kept on auditing problematic firms, and they got driven out of business. Were they unusual in their seeming lack of commitment to real monitoring and in their failure to cut off clients who were predicted to behave poorly in the near future? Not in my experience. The lack of commitment to effective monitoring is a major concern in game-theory designs for organizations. This is true because, as we will see, too often companies have weak incentives to know about problems. Was the lack of monitoring rational? Alas, yes, it was, even though in the end it meant the demise of Arthur Andersen, LLP. Game-theory thinking made it clear to me that Andersen would not monitor well, but I must say Andersen's most senior management partners genuinely did not seem to understand the risks they were taking.

At Arthur Andersen, partners had to retire by age sixty-two. Many retired at age fifty-seven. These two numbers go a long way toward explaining why there were weak incentives to pay attention to audit risks. The biggest auditing gigs were brought in by senior engagement partners who had been around for a long time. As I pointed out to one of Andersen's senior management partners, senior engagement partners had an incentive not to look too closely at the risks associated with big clients. A retiring partner's pension depended on how much revenue he brought in over the years. The audit of a big firm, like Enron, typically involved millions of dollars. It was clear to me why a partner might look the other way, choosing not to check too closely whether the firm had created a big risk of litigation down the road.

Suppose the partner were in his early to mid-fifties at the time of the audit. If the fraud model predicted fraud two years later, the partner understood that meant a high risk of fraud and therefore a high risk that Andersen (or any accounting firm doing the audit) would face costly litigation. The costs of litigation came out of the annual funds otherwise available as earnings for the partners. Of course, this cost was not borne until a lawsuit was filed, lawyers hired, and the process of defense got under way. An audit client cooking its books typically was not accused of fraud until about three years after the alleged act. This would be about five years after the model predicted (two years in advance) that fraud was likely. Costly litigation would follow quickly on the allegation of fraud, but it would not be settled for probably another five to eight years, or about ten or so years after the initial prediction of risk. By then, the engagement partner who brought in the business in his early to mid-fifties was retired and enjoying the benefits of his pension. By not knowing the predicted risk ten or fifteen years earlier, the partner ensured that he did not knowingly audit unsavory firms. Therefore, when litigation got under way, the partner was not likely to be held personally accountable by plaintiffs or the courts. Andersen (or whichever accounting firm did the audit) would be held accountable (or at least be alleged to be accountable), as they had deep pockets and were natural targets for litigation, but then the money for the defense was coming out of the pockets of future partners, not the partner involved in the audit of fraudulent books a decade or so earlier. The financial incentive to know was weak indeed.

When I suggested to a senior Andersen management partner that this perverse incentive system was at work, he thought I was crazy—and told me so. He thought that clients later accused of fraud must have been audited by inexperienced junior partners, not senior partners near retirement. I asked him to look up the data. One thing accounting firms are good at is keeping track of data. That is their business. Sure enough, to his genuine shock, he found that big litigations were often tied to audits overseen by senior partners. I bet that was true at every big accounting firm, and I bet it is still true today. So now we can see, as he saw, why a partner might not want to know that he was about to audit a firm that was likely to cook its books.

Why didn't the senior management partners already know these facts? The data were there to be examined. If they had thought about incentives more carefully, maybe they would have saved the partnership from costly lawsuits such as those associated with Enron, Sunbeam, and many other big alleged frauds. Of course, they were not in the game-theory business, and so they didn't think as hard as they could have about the wrong-headed incentives designed into their partnership (and most other partnerships, for that matter).

On the plus side, management's incentives were better than those of the engagement partners. Senior managers seemed more concerned about the long-term performance of the firm. Maybe that was the result of what we call a selection effect, as people concerned about the firm's well-being may have been more likely candidates to become senior management. Still, they also had an incentive to help their colleagues bring in business, and that meant that they were interested in making it easy for their colleagues to sign up as many audit engagements as possible. They may have preferred to avoid problems with bad clients, but the senior managers could live with not knowing about future trouble if that helped to keep their colleagues happy and business pouring in. Thus, senior management's incentives were not quite right either. Effective monitoring had benefits for them, but it was costly in revenue and especially in personal relations. Many senior management partners tolerated slack monitoring as the solution to this problem, and likely did a quick risk calculation that litigation—not collapse—was the worst that a fraudulent client could visit upon the firm. Let's face it, many of us would do the same thing.

We also should remember that but for what seems to have been an overly zealous prosecution by the Department of Justice, the likely risk calculation by senior management partners would have been right. Remember, while Andersen gave up its license to engage in accountancy in 2002, following its conviction on criminal charges, the conviction was overturned by the Supreme Court. Sadly for the approximately 85,000 people who lost their jobs, the Supreme Court decision came too late to save the business.

It's hard for anyone to enforce policies that day in and day out tick off colleagues. That's especially true if these colleagues are the ones who choose which partners will get to be the senior managing partners. In partnerships like Arthur Andersen or any of the other big accounting firms (or law firms), the people who run the organization are elected by their colleagues. Their engagement partners, not the senior managers, are the rainmakers who keep money pouring in.

The perverse incentive structure that discourages companies from accurately anticipating fraud is not unique to the accounting business. We can see the same problems in the insurance and banking industries. Suppose, for instance, you told underwriters to stop selling directors' and officers' insurance to a big client like Enron in 1995. In 2001 the SEC alleged that Enron had committed securities fraud starting around 1997 or 1998. Before that, Enron was a well-regarded company. During all of those years between 1995 and 2001, your colleagues, the insurance underwriters, would be screaming that you were taking their income away, that there was no evidence that Enron was doing anything wrong, that in fact it was a fine and prosperous company. In their eyes, you were giving their business away to people working for rival firms. That's a pretty tough case to refute while you wait five, ten, or fifteen years for the other shoe to drop. You can imagine how hard it must be to get a real commitment to monitor and punish misconduct, since one must be careful, of course, not to jump in and punish employees or clients before you are sure they have done something wrong. There are big costs attached to falsely accusing a client of fraud, just as there are big costs attached to incorrectly trusting that a firm is behaving honestly.

Management can be a profile in courage by cutting off revenues today to prevent bigger headaches tomorrow, but most profiles in courage, as it turns out, lose their jobs. That's not an easy choice for anyone. Sure, we all pay lip service to the idea that we should do what is good for us and our colleagues in the long run, but doing what is good in the long run can be very costly in the near term. As Lord Keynes so aptly observed, in the long run we're all dead (or, anyway, retired). Losing business now to avoid lawsuits later is hard for exactly that reason.

As we've explored, game theory predicts that people frequently, for rational reasons, assume great risk and experience great failure. I suppose you could say that making predictions for a living makes that very possibility something of a daily routine. Thankfully, my record has been pretty good, but there have been some notable misses. And indeed there are some other associated risks with the further refinement of rational choice theory and the models I develop and employ. The next chapter will examine some of these issues.

8

■

HOW TO PREDICT
THE UNPREDICTABLE

THIS CHAPTER IS about the limitations of my models, some of
my worst-ever predictions, and some of the potential dangers that
can conceivably stem from "predictioneering." Many a critic of mine will
have well-worn and dog-eared pages in this stretch of the book!

My worst-ever prediction came in the months after Bill Clinton's election to the presidency. When he was elected, it was obvious to everyone
that he was going to try to push through a comprehensive health care
plan. He assigned his wife to head a task force charged with designing a
health program. At the time, I was engaged by a major brokerage firm to
help work out what was likely to get through Congress so that they could
design investment opportunities around the new program. As we all know
now, the task force created a lot of heat, but no agreement on a new
health care program. Instead, it failed dismally.

As it happened, my analysis of what the health care plan would look
like led to one of my worst-ever predictions. Each and every detail of what
came out of my analysis was both wrong and filled with lessons that improved future assessments. Models fail for three main reasons: the logic
fails to capture what actually goes on in people's heads when they make
choices; the information going into the model is wrong—garbage in,
garbage out; or something outside the frame of reference of the model oc-
curs to alter the situation, throwing it off course. The last of these is what
happened to my health care analysis.

In early 1993, I predicted what was likely to get through Congress sometime in 1993 or 1994. In some sense, all three of the limitations I mentioned were involved and were subsequently addressed as part of my personal learning experience. But by and large, the main problem had to do with an unforeseen event that completely altered the setting in which health care was going to be shepherded through Congress. Of course, the whole point of prediction is to forecast the unforeseen. Anyone can predict that the sun will rise in the east and set in the west tomorrow. Still, there's unforeseen and then there's *unforeseen*. I think you'll see what I mean when we go through what happened to health care, at least as I looked at it.

Although the experts who provided the data identified a great many components of a comprehensive health plan—including questions related to long-term care, proportion of the population covered, costs of drugs, distribution of the tax burden for health care across the federal and state governments, as well as employers' costs, total spending on health care, and even questions related to ancillary care—that would get congressional approval, none did. As it happens, the model predicted that Daniel Rostenkowski, then an influential Illinois congressman and, crucially, chairman of the powerful House Ways and Means Committee, was the key to getting health care legislation through Congress. Mr. Rostenkowski, however, was indicted on seventeen felony counts of corruption in 1994 (and later convicted) based on investigations that reached their height during 1993, as the Clinton White House's health care push began in earnest. Rostenkowski's salience for health care plummeted, of course, first in anticipation of his indictment and then even more as he fought to salvage his reputation, maintain his leadership position in Congress, and keep himself out of prison. He failed on all counts, and my prediction, based on his effective efforts on behalf of health care, also failed. As a result, contrary to my expectations, nothing passed through Congress.

Rostenkowski's indictment was a shattering shock to the situation as analyzed; I'll explain why in a moment. The model assumed, incorrectly, that the underlying conditions would remain unaltered during the period of negotiation and bargaining over health care. My client was not terribly happy that I got everything wrong, and neither was I, but at least I had the benefit of learning an important lesson. It was little consolation to know that if I repeated my analyses after dropping Rostenkowski from the data set, I got everything right. Without Rostenkowski, the model showed that agreement would not be reached in the House of Representatives, and that meant there would be no comprehensive health care plan. But, of course, that was analysis done in hindsight, and that is no way to help a client. Needless to say, the client was not particularly understanding or forgiving and never asked me to do another piece of work—a great disappointment, because I would have welcomed the opportunity to prove to them the value of modeling, and to do so for free. But they didn't bite, and who can blame them? They invested valuable time as well as money in my analysis, and they had absolutely nothing to show for it.

What did my study find and why did it find it? The Rostenkowski study, as I now think of it, had a long list of players that included several members of Congress, Hillary Clinton herself, health care expert advisers from nursing homes, AARP, pharmaceutical companies, employers of all shapes and sizes, and so forth. Many issues were relatively difficult to resolve within the model's own logic, taking many rounds of negotiation, posturing, and information exchanges before settling on what looked like a stable outcome—that is, an outcome that could get through the House and Senate. It was evident that more compromise was needed than some key players were prepared to accept. It was also evident that the study had to involve at least two (and possibly as many as four) distinct phases.

The first phase, common in many analyses of legislative decisions, focused on the period of lobbying and jockeying for position. In this phase, all of the players with an interest in shaping the outcome are part of the analysis. That includes many stakeholders who would not have a place at the table when it came time for the House and Senate to vote and for the president to sign or veto whatever they sent up to him. Organizations like Blue Cross–Blue Shield or the AMA that were utterly opposed to the Clinton plan, or some labor union leaders and local government interests that were strongly in favor of the plan, are included in the lobbying phase along with the decision makers. Then, when the lobbying game ends (according to the model's rules), the analysis moves to the next phase. Because of the pulls and tugs during the lobbying period, many players' positions will have shifted. They will have responded to offers of compromise or to coercion or to the anticipation of such pressures. So at the end of that first game, the decision makers move on to the next phase, but not with their original positions on individual health care issues intact. They move on at whatever position the model predicts they will hold when the lobbying game ends.

The next phase then pits just the decision makers against each other. Gone are labor union leaders, the AMA, the media, the Blues, and local and state governments, and gone is Hillary Rodham Clinton. Sure, she had influence in the lobbying phase, but she didn't get to vote in Congress. From the model's perspective, whatever whispers there might have been between her and President Clinton ended with the lobbying phase. He, and others, had ample opportunity in the first phase to succumb to, adjust to, or resist her arguments.

The second phase predicted the passage of a comprehensive bill in both the House and the Senate. It also predicted that the bill that would come before President Clinton was one he could easily have signed, although it would have been much altered from the legislation sought by Hillary Clinton. So there was little need in this case to do a further analysis to work out the negotiations between the House and Senate leadership over the exact contents of the proposed legislation, and there was no need to do a detailed study of the risks of veto and the prospects of overriding a veto. It just wasn't an issue.

The numbers having been crunched, four results popped out of the analysis as being crucial to understanding where health care reform was headed. First, Hillary Clinton was an unusual stakeholder, not because she was First Lady, but because she showed the characteristics of someone who was content to fail while sticking to her principles. Despite pressure on her from every side, she was nearly immovable on each and every one of the issues I examined. This is a characteristic that is rarely seen in democratic politics (although many perceived this as the bargaining approach, really a bullying approach, used by George W. Bush). Sure, I had seen a rigid adherence to positions before in other studies I had done. The late Nigerian general Sani Abacha (and I do not mean to compare Hillary Clinton or George W. Bush to him on any substantive grounds—just bargaining style back then) was an important focus of many studies I did. He hardly ever shifted position, but then he didn't have to. He got to dictate outcomes. Hillary Clinton barely shifted positions either, but, from a practical standpoint, she needed to. All the evidence suggests that her time in the U.S. Senate after her husband's presidency made her a shrewd judge of when to flex some muscle and when to muster some flexibility. That should serve her well in the world now, but back then the model said she only knew about flexing muscle.

In the language of the times, Hillary Clinton had a tin ear when it came to politics. Fair enough. She had never run for office and had not been a politician. But her rigid willingness to go down in a seeming blaze of glory, but inevitably down nevertheless, hurt the chances of forging compromises with those who saw themselves excluded from the debate. This includes such important interest groups as the American Medical Association, much of the pharmaceutical industry, and others from whom even grudging support would have made selling health care reform much easier. Indeed, the analysis suggested that given the right response from the Clinton health care task force, the AMA was more flexible on many health care issues than was commonly perceived at the time. They could have been brought around to support a program that could have passed the House and Senate.

The second striking result was Bill Clinton's bargaining style within the model's logic. There are two ways to maneuver into a winning position. One is to persuade others to adopt your point of view. The other is to adopt theirs. Bill Clinton—in the model's logic; I don't know what he was actually doing behind closed doors—was the latter type. This was probably due to the modest degree of salience coupled with a fairly centrist, even slightly right position assigned to him on most health care issues by the expert panel I used to create the inputs for the model.

As the model saw things, President Clinton would sniff out where the strongest coalition was, and he would move close to it. He was like a person with a wet finger stuck in the wind to see which way the breeze is blowing. If Hillary Clinton's principle at the time can be described as "Back what you believe in, come hell or high water," Bill Clinton's principle was "Win, no matter what constitutes winning." With the benefit of hindsight, writing this a decade and a half later, I think it fits pretty well with what many have come to think of as Bill Clinton's governing style.

The third striking result was how ineffective many members of Hillary Clinton's task force were at looking past their personal beliefs. They were more open to compromise than Hillary Clinton, but they were reluctant to take her on, and so they were willing to accommodate the opposition too little to build a bridgehead to victory. In the model's vocabulary, they gave in to her while failing to realize that they had more potential to change her mind than they thought.

And the fourth result, the really striking result, was that Dan

Rostenkowski—that is, the person controlling the purse strings in the purse-string-shaping House Ways and Means Committee—had none of these limitations. He maneuvered skillfully—again, I am talking in terms of the model's predictions; I don't know what really went on, I only know how it turned out. He knew how to alter the thinking of other players from Congress. He knew how to reshape the president's thinking and the perspective of many on the task force. He also handled the opposition lobbyists and interest groups skillfully.

What did the model see in Rostenkowski that was absent from the Clintons or other players? In order for health care reform to matter, it had to be funded, of course, and that was the province in which Dan Rostenkowski exerted the greatest influence. The expert data, not surprisingly, rated him as enormously powerful on questions related to how to pay for health care reform. And he was moderately conservative on this question, wanting to shift most of the cost away from the federal budget. Bill Clinton was perceived to argue for an even more conservative position when it came to paying for health care. So Rostenkowski was seen as the more moderate of the two, and he was believed to be as powerful as the president on this question.

Rostenkowski was positioned at a point that had a great mountain of powerful support behind it, he had enough salience to influence people (but not so much as to come across as excessively intense and committed), and he was surrounded by small, fragmented clusters of influence scattered across many positions with relatively little clout to withstand his pressure. In that environment (according to the model's logic), Rostenkowski was positioned as a leader who could and would move people to his position. He found the right arguments and the right opportunities and the right targets to cajole or coerce so that the prospective winning position was located close to where he wanted it to be. He didn't go to the winning position, he brought it to him. Thus, on one health care issue after another, because he exerted so much control over the money, what Rostenkowski wanted, Rostenkowski could pretty much get. Except, ah, except for those seventeen felony counts. They were not part of my analysis, they were truly *unforeseen*, and they made all of the difference. The felony counts were exogenous shocks—that is, the product of outside, unexamined forces unrelated to health care issues.

The political world and the business world are vulnerable to unanticipated shocks. With the Rostenkowski experience in hand, I realized I needed to have a way to anticipate unpredictable events so that I could take them into account. But how can you predict the unpredictable? Well, although it is impossible to anticipate unpredictable developments, it is possible to predict how big an "earthquake" is needed to disrupt a prediction. I worked out how to predict the magnitude of disruptions even if I could not know their exact source. We'll look at how I have since incorporated this element into my work.

How I got to the ever-evolving solution is an interesting story in itself. Around the time that Dan Rostenkowski's troubles led me to think about random shocks, John Lewis Gaddis, a world-renowned historian, now at Yale University but then at the University of Ohio in Athens, Ohio, invited me to spend a week with him and his students. Gaddis had written a paper in 1992 claiming that international relations theory was a failure because it didn't predict the 1991 Gulf War, the demise of the Soviet Union, or the end of the cold war. Two well-known political scientists, Bruce Russett at Yale University and James Ray at Vanderbilt University, responded that Gaddis had not taken my predictive rational-choice work into account.[1] They contended that this body of work warranted being viewed as a rigorous scientific theory rather than just some exercise in fitting data to outcomes after they were known.

Gaddis paid attention to the claim by Russett and Ray that he had overlooked a relevant body of research. That is what led him to invite me to spend time with him and his students. He was a doubter, and he made no bones about that. Southern gentleman that he is, he couched his doubts in the most civil way; still, I was going to Athens, Ohio, with the expectation on John's part that his students and he were going to expose my modeling as some sort of hocus-pocus.

I agreed to apply my method to any policy problem that Gaddis and his students agreed on, although I imposed two restrictions. First, they had to know enough about the issue they chose to be able to provide me with the data needed for the model, since I was unlikely to be an expert on the issue and anyway it would be best if the data came from doubters. Second, it had to be an issue for which the outcome would be known over a period ranging from a few months to a year or two, rather than something that would not be known for so long that they couldn't judge in a timely fashion whether my model's logic had gotten it right or not. The hope was to look at something we could then correspond about. They would know I had made predictions before the fact, and, the timing being right, they would be able to look back at what I had said and compare it to what actually happened later.

They chose to have me analyze what became the 1994 baseball strike. I made detailed predictions about such matters as whether there would be a strike (the model said yes), whether there would be a World Series that year (the model said no), and whether President Clinton's eventual threatened intervention, which was predicted by the model, would end the strike (again, the prediction was no). I did an in-class interview with the two or three students who were "experts" on baseball, and then ran my computer model in front of all the students. I provided an analysis of the results on the spot. That way the students knew that nothing more went into my predictions than the data collected in class and the logic of my model. The predictions, as it happened, turned out to be correct.

Shortly before I left Athens, Professor Gaddis suggested that I write a paper applying the model to the end of the cold war. In particular, he proposed that I investigate whether the model would have correctly predicted the U.S. victory in the cold war based only on information that decision makers could have known shortly after World War II ended. That is, he asked for a sort of out-of-sample prediction of the type I used to validate the fraud model. And so my analytic experiences with Dan Rostenkowski and with the baseball strike came together to provide a motivation and a framework for assessing the end of the cold war. My work on this project, using only information available in 1948, would help me incorporate and test my new design for external shocks within a model, which, of course, I felt compelled to develop on account of my health care debacle.

I used Gaddis's proposal and my awful experience with health care to think through how to predict the consequences of inherently unpredictable events. I put together a data set that my model could use to investigate alternative paths to the end—or continuation—of the cold war. The data on stakeholder positions were based on a measure of the degree to which each country in the world as of 1948 shared security interests with the United States or the Soviet Union. The procedure I used to evaluate shared interests was based on a method I developed in publications in the mid-1970s.[2] The procedure looks at how similar each pair of countries' military alliance portfolios are to each other from year to year. Those who tended to ally with the same states in the same way were taken to share security concerns, and those who allied in significantly different ways (as the United States and the Soviet Union did) were taken to have different, perhaps opposed, security policies and interests.

The correlation of alliance patterns as of 1948 was combined with information on the relative clout or influence of each state in 1948. To asses clout I used a standard body of data developed by a project then housed at the University of Michigan called the Correlates of War Project. Those data, like my measure of security interests, can be downloaded by anyone who cares to. They are housed at a website called EUGene, designed by two political science professors who were interested in replicating some of my research on war.[3]

Each state in the data set—I focused on countries rather than individual decision makers to keep the data simple and easy to reproduce by others—was assigned a maximum salience score to reflect the urgency of security questions right after the Second World War. Combining these data to estimate expected gains and losses from shifting security policies, the model was run on these data for all country-pair combinations one hundred times. Each such run consisted of fifty "bargaining periods." The "bargaining periods" were treated as years, and thus the model was being used to predict what would happen in the cold war roughly from 1948 until the end of the millennium.

Each country's salience score was assigned a one-in-four chance of randomly changing each year. That seemed high enough to me to capture the pace at which a government's attention might move markedly in one direction or another and not so high as to introduce more volatility than was likely within countries or across countries over relatively short intervals. Naturally, this could have been done with a higher or lower probability, so there is nothing more than a personal judgment behind the choice of a one-in-four chance of a "shock."

Any changes in salience reflected hypothetical shifts in the degree to which security concerns dominated policy formation or the degree to which other issues, such as domestic matters, surfaced to shape decision making for this or that country. Thus, the salience data were "shocked" to capture the range and magnitude of possible political "earthquakes" that could have arisen after 1948. This was the innovation to my model that resulted from the combination of my visit to Ohio and my failed predictions regarding health care. Since then, I have incorporated ways to randomly alter not only salience but also the indicators for potential clout and for positions, and even for whether a stakeholder stays in the game or drops out, in a new model I am developing.

Neither the alliance-portfolio data used to measure the degree of shared foreign interests nor the influence data were updated to take real events after 1948 into account. The alliance-portfolio measure only changed in response to the model's logic and its dynamics, given randomly shocked salience. Changes in the alliance correlations for all of the countries were the indicator of whether the Soviets or the Americans would prevail or whether they would remain locked in an ongoing struggle for supremacy in the world.

So here was an analysis designed to predict the unpredictable—that is, the ebb and flow of attentiveness to security policy as the premier issue in the politics of each state in my study. With enough repetitions (at the time, I did just a hundred, because computation took a very long time; today I would probably do a thousand or more) with randomly distributed shocks, we should have been able to see the range of possible developments on the security front. That, in turn, should have made it possible to predict the relative likelihood of three possible evolutions of the cold war: (a) it would end with a clear victory by the United States within the fifty-year period I simulated; (b) it would end with a clear victory by the Soviet Union in that same time period; or (c) it would continue, with neither the Soviet Union nor the United States in a position to declare victory.

What did I find? The model indicated that in 78 percent of the scenarios in which salience scores were randomly shocked, the United States won the cold war peacefully, sometimes by the early to mid-1950s, more often in periods corresponding to the late 1980s or early 1990s. In 11 percent of the simulations, the Soviets won the cold war, and in the remaining 11 percent, the cold war persisted beyond the time frame covered by my investigation. What I found, in short, was that the configuration of policy interests in 1948 already presaged an American victory over the Soviet Union. It was, as Gaddis put it, an emergent property. This was true even though the starting date, 1948, predated the formation of either NATO or the Warsaw Pact, each of which emerged in almost every simulation as the nations' positions shifted from round to round according to the model's logic.[4]

The selection of 1948 as the starting date was particularly challenging in that this was a time when there was concern that many countries in Western Europe would become socialist. This was a time, too, when many thought that a victory of communism over capitalism and authoritarianism over democracy was a historical inevitability. On the engineering front it was, of course, too late to change the course of events. Still, the model was quite provocative on this dimension, as it suggested opportunities that were passed up to win the cold war earlier. One of those opportunities, at the time of Stalin's death (which, of course, was not a piece of information incorporated into the data that went into the model), was, as it turns out, contemplated by real decision makers at the time. They thought there might be a chance to wrest the Soviet Union's Eastern European allies into the Western European fold. My model agreed. American decision makers did not pursue this possibility, because they feared it would lead to a war with the Soviet Union. My model disagreed, predicting that the Soviets in this period would be too preoccupied with domestic issues and would, undoubtedly with much regret, watch more or less helplessly as their Eastern European empire drifted away. We will, of course, never know who was right. We do know that that is what they did a few decades later, between 1989 and 1991.

So with the help of Dan Rostenkowski and John Gaddis's students I was able to show how strongly the odds favored an American cold war victory. The account of the cold war, like the earlier examination of fraud, reminds us that prediction can look backward almost as fruitfully as it can look forward. Not everyone was as generous as John Gaddis in acknowledging that game-theory modeling might help sort out important issues, and not everyone should be (not that it isn't nice when people are that generous). There should be and always will be critics.

There are plenty of good reasons for rejecting modeling efforts, or at least being skeptical of them, and plenty of bad reasons too. Along with technical failures within my models, or any models for that matter, there is the obvious limitation in that they are simply models, which are, of course, not reality. They are a simplified glance at reality. They can only be evaluated by a careful examination of what general propositions follow from their logic and an evaluation of how well reality corresponds with those propositions. Unfortunately, sometimes people look at lots of equations and think, "Real people cannot possibly make these complicated calculations, so obviously

real people do not think this way." I hear this argument just about every se-
mester in one or another course that I teach. I always respond by saying that
the opposite is true. Real people may not be able to do the cumbersome
math that goes into a model, but that doesn't mean they aren't making much
more complicated calculations in their heads even if they don't know how
to represent their analytic thought processes mathematically.

Try showing a tennis pro the equations that represent hitting a ball with
topspin to the far corner of the opponent's side of the court, making sure
that the ball lands just barely inside the line and that it travels, say, at 90
miles an hour. Surely the tennis pro will look at the equations in utter be-
wilderment. Yet professional tennis players act as if they make these very
calculations whenever they try to make the shot I just described. If the
pro is a ranked player, then most of the time the shot is made successfully
even though the decisions about arm speed, foot position, angle of the
racket's head, and so forth must be made in a fraction of a second and
must be made while also working out the velocity, angle, and spin of the
ball coming his or her way from across the court.

Since models are simplified representations of reality, they always have
room for improvement. There is always a trade-off between adding complexity and keeping things manageable. Adding complexity is only warranted when the improvement in accuracy and reliability is greater than the
cost of adding assumptions. This is, of course, the well-known principle of
parsimony. I've made small and big improvements in my game-theory modeling over the years. My original forecasting model was static. It reported
what would happen in one exchange of information on an issue. As such, it
was a good forecaster but not much good at engineering. While I was tweaking that static model to improve its estimation of people's willingness to take
risks and to estimate their probability of prevailing or losing in head-to-head
contests, I was also thinking about how to make the process dynamic. Real
people, after all, are dynamic. They change their minds, they switch positions on questions, they make deals, and, of course, they bluff and renege
on promises.

About ten years after creating the static version I finally worked out a
dynamic model I was happy with. That is the model I'm mostly discussing
in this book. Over the past few years I've been working on a completely
new approach based on a more nuanced game than the one I described
back in the third chapter. Preliminary tests of this new model indicate

that it not only yields more accurate predictions, but also captures play dynamics more faithfully. As an added bonus, it also allows me to evaluate trade-offs across issues or across different dimensions on a single issue simultaneously. It also gives me the opportunity to assess how each player's salience and influence changes from bargaining round to bargaining round, something the older model cannot do. I will apply this new model to some ongoing foreign policy crises and to global warming in the last two chapters. That will be my first foray into opening the opportunity to be embarrassed by my new approach.

The process of discovery is never-ending. That's both the challenge and the excitement behind doing this kind of research: finding better and better ways to help people solve real problems through logic and evidence. Not everyone, though, shares my enthusiasm for this sort of effort at discovery.

Some critics object to predicting human behavior. They worry that government or corporations will misuse this knowledge. They're concerned about the ethics of reducing people to equations. To me this is an odd set of objections, especially since it comes mostly from people who are unhappy with the quality of government policy choices and with corporate actions to begin with. Some of my academic colleagues particularly object to providing guidance to the intelligence community, the "evil" CIA, on national security matters. They seem to think that the government shouldn't have the best tools at its disposal to make the best choices possible. I don't share that view. If we want better decisions from our government, we ought to be willing to help it improve its decision making.

Yes, there is always a risk that any tool will be misused. But science is about understanding how the world works. Different people have different personal views about what will make the world a better place, and it's the job of officials and citizens to regulate unethical uses of information. Further, it is the responsibility of each of us as individuals to withhold our expertise when we think its use will make the world, or our little part of it, a worse place.

■ ■ ■

I turn down clients when I don't want to help them achieve their goals. Many years ago, for example, I was approached by someone claiming to represent the Libyan government. The person who contacted me wanted me to figure out how to facilitate overthrowing the Egyptian government then led by Anwar Sadat. The contact proposed flying me to Geneva, Switzerland, to avoid the possibility of the United States government or some other government being able to subpoena the results of my then very primitive modeling effort. I was offered a million dollars for my trouble. There is no way for me to know whether this approach was authentic or a hoax, although it certainly seemed real. I declined and immediately contacted people in the U.S. government to alert them to my experience.

Several years later I was contacted by yet another person with an unsavory proposal. This person represented himself as an agent for Mobutu Sese Seko of Zaire. Mobutu's hold on power had become tenuous. His economy was doing poorly, his soldiers were becoming agitated, and his loyal followers were becoming shaky because he was known to have a terminal illness. They were presumably worried about who would protect them and take care of them financially when he was gone. The contact person wanted to know if I could work out how to salvage Mobutu's control over Zaire and offered a success fee of 10 percent of Mobutu's offshore financial holdings. I know this sounds hard to believe, but it happened, and it was before unscrupulous Nigerians had worked out their famous Internet bank scams.

Mobutu at the time was reputed to be worth somewhere between $6 billion and $20 billion. If this had been for real and if I could have engineered his continuation in office until he died peacefully or chose to step down, and if I had been willing to do so, I could have been paid an unbelievable fortune. But even if the fortune had been believable, the answer would still have been the same. As in the alleged Libya offer, I said no without a moment's hesitation. I was confident that Mobutu's difficulty was an analytic problem with a solution, but no amount of money could have justified my intervention. My main concern was that I would be on the radar screen of people I really preferred not to know about me. And once again I contacted people in the U.S. government to alert them to the situation.

Of course, my personal judgment about who to do business with might differ from someone else's judgment. I couldn't see any justification for helping anyone topple Sadat. Here was a man who had put his life at risk—and would tragically lose it—in a sincere and successful effort to advance the cause of peace. Mobutu's case could be seen as (ever so slightly) more complicated. There was a slender ethical case to be made on Mobutu's behalf. While it did not appeal to me, one could easily have argued that whoever came after Mobutu might be even worse. Back then, and even immediately after his overthrow, it wasn't clear that the Congo was moving in a better direction. Still, for me the answer was unambiguous. For others—who knows how they would have evaluated the pros and cons of applying insights from science to help or hinder a dictator like Mobutu?

Some of you may think I should not use game theory to help big corporations get good settlements in litigation, especially when their opponents in civil matters may not be able to afford (or choose not to afford) comparable help. Others may think I don't do enough to help plaintiffs (although my firm is happy to do so; we just aren't asked very often), or what have you. Still others may subscribe to the lawyer's dictum: Everyone is entitled to the best defense they can muster. We all have our individual standards about how to use or withhold our knowledge and skills, and that is as it should be.

In the end, I believe advances in scientific knowledge almost always better the human condition. If we turn ourselves into Luddites, we'll just shift the advantages of knowledge to others. Remember, after Galileo's persecution by the Catholic Church, physics went into decline in Italy for centuries, until, perhaps, the arrival of Enrico Fermi on the scene. Despite the setbacks in Italy, that didn't mean research into physics stopped. It moved to Protestant northern Europe, leaving Italy to fall behind. Similarly, efforts to stymie science in China caused that country, once the world's most advanced in scientific knowledge and discovery, to descend into scientific oblivion. China's emperors chose to have their people look within themselves rather than at the stars; China is still struggling to overcome the deficit it created for itself. I hope we will not make the same mistake. As for me, I continue to look for ways to improve my understanding of how the world of strategic human behavior works. And that's central to my motivation to continue learning from past failures.

As I said earlier, and as we've seen in this chapter, prediction can look backward almost as fruitfully as it can look forward, providing remarkable insight not only into what happened but also into what might have been. Accordingly, in the next chapter we'll have some fun with history. We will look at how World Wars I and II might have been avoided, and how Sparta might have prevented its colossal collapse after its stunning victory in the Peloponnesian War. And while everyone knows Columbus sailed the ocean blue in fourteen hundred and ninety-two, what they don't know is that his experience presents an interesting bargaining problem—one whose outcome explains why Spain said yes and why Portugal (among others) said no, forever changing the course of history. In looking at the past with a game-theory microscope we begin to grasp the logic behind the history we know (and a sense of just how *un*-inevitable history is) and, sometimes to tragic effect, the missed opportunities for strategic choices that would have altered its course.

9

■

FUN WITH THE PAST

ERE ARE QUESTIONS and brief answers about four really big
events in history:

Why did Sparta lose its hegemonic position in Greece just thirty-three
years after victory in the Peloponnesian War?
Because Spartans loved their horses more than their country.

Why did Ferdinand and Isabella decide to fund Columbus?
Because he agreed to work cheap.

How could World War I have been avoided?
By British sailors taking a summer cruise to the Adriatic.

How could World War II have been avoided?
By German Social Democrats making nice to the pope.

If the Spartans had not been so fond of horse races, we might all speak
Greek today. If the British had been a little more adept at diplomacy in 1914,
Austrians and Germans might speak English today, which, come to think of
it, many of them do. Maybe in that case there would never have been a Russian Revolution, maybe Adolf Hitler would have stuck to painting, maybe
there would not have been a Second World War or cold war or Winston
Churchill (at least as we know him), and maybe the sun still would not set

on the British Empire. We will never know. But we can approximate what might have happened if Sparta's horses had run fewer races, or if Britain's diplomats had shipped some of their sailors up the Adriatic in 1914, or if German socialists had shown more flexibility toward Catholics in 1933.

Let's be fair to the decision makers of the past. Just like those of today, they were hampered by difficult choices, complicated incentives, and poor foresight. Of course they could have done better if they had had a stealth bomber or a nuclear deterrent or a high-speed computer, but they didn't. Does that mean their hands were tied? In one sense, yes, it does. They knew what they knew and did not have the technological or scientific foundation to do much better. But in another sense, we should not underestimate what they might have done. They did have logic, and, let's face it, logic is logic. Its fundamentals have not changed in millennia. With enough people sitting around banging out calculations with an abacus or writing down results in the sand, they might have thought up and solved a model like mine, or one that's better.

I will explore all of these questions and their answers in this chapter. To do so, however, I have to explain a little bit more about how to think realistically about altering the world. Sure we can play games like "What if Napoleon had had a stealth bomber at Waterloo?" (probably not as useful an advantage as a few machine guns)—but he didn't, and couldn't. I prefer to play realistic games. That's what we will do here. We will ask how some big events might have turned out differently if realistic alternative strategies had been pursued. So, how to think about what might have been? Answer: Think about what people could have done but chose not to do, and why.

An acquaintance of mine studies the history of religion in a mostly political context. He is especially interested in the history of religion in Russia, and especially in its survival despite seventy years of official state-sponsored atheism. He once pointed out to me, both in bemusement and with amusement, that the way I differ from a historian is that I spend 95 percent of my time thinking about what *didn't* happen. He is probably right. From the perspective of many historians, how things ended up was either inevitable, or, in a diametrically opposed way, the result of chance that could have swung any which way.

I am not big on the idea of historical inevitability. If that were right, there would be no point in trying to be a political engineer, a predictioneer. The idea that history is a play and that we are acting out scripted parts with little freedom of choice seems silly at best and downright evil at worst. It risks justifying anything anyone does, no matter how terrible. This view says, "Blame the writer, not the actor." I won't venture to guess who the writer might be.

Conversely, the notion that the developments that make up history are primarily a series of chance events seems equally odd to me. Why fight over ideas, select governments, build armies, fund research, promote literacy, create art, or write histories if all we are doing is twiddling our thumbs while chance developments send us bouncing around like the physicist's particles? How can anyone deny strategic behavior and its consequences when we are surrounded by it in almost everything we do?

To be sure, the world as we know it could have swung one way or the other. That's why neither the past nor the future follows an inevitable path. There are always chance elements behind which ways things swing, but those chance events rarely decide the future. Bad weather may have been important in Germany's failed invasion of Russia during World War II, but Hitler's choice to delay his invasion while turning his attention to problems in Yugoslavia was a calculated risk. He regretted it later, but still, he knew delay increased the odds that the German army would face bad weather.

The December 23, 2006, 6.7 earthquake in Bam, Iran, certainly was outside anyone's control. It cost more than 26,000 lives. That represented nearly 20 percent of the approximately 142,000 residents of the area. Interestingly, just days before, a 6.5 earthquake shook the Southern California town of Cambria. Three people died out of the nearly 250,000 in the surrounding area. The Loma Prieta earthquake in 1989 killed 68 people in the San Francisco–Oakland area, a metropolitan area with a population of more than five million. The 1989 temblor was about five times larger than the Bam quake, and yet it killed—not to minimize the tragedy—only a tiny fraction of the local population, while in Bam the death toll was horrendous. Is the vast discrepancy between deaths in California and deaths in Iran from earthquakes a matter of chance, was it inevitable, or was it a matter of strategic decisions?

The answer the press offered at the time was that people in Bam lived in mud and stone houses and people in California did not. Yet we must ask:

Why do people in a wealthy country with vast oil reserves live in mud houses? It is tempting to speak of natural disasters—earthquakes, floods, droughts, famines—as if some parts of the world just have terribly bad luck. That these are dreadful natural events there is no doubt, but are they really *natural* disasters? Certainly such terrible events are random from a political or social perspective. Their causes are well beyond human control, at least given the current state of knowledge in predicting earthquakes, hurricanes, droughts, and tsunamis. But their consequences are not.

Death tolls from cataclysmic natural events are vastly higher in countries run by dictators than in democracies. Democratic governments prepare for disasters, regulate construction to increase the chances of surviving events like earthquakes, and stockpile food, clothing, and shelter for disaster victims. Why? Because governments elected by the people are largely accountable to the people. Governments selected by the military, or the aristocracy, or the clergy, or the one legal political party are accountable to very few. It is those few they protect, not the many, because it is at the pleasure of those few that they rule. No, even chance events rarely have consequences that are primarily due to chance. Strategic choices lurk behind who wins and who loses, who lives and who dies, even when nature seems to be the prime culprit.

Let's take a look at some important turning points in history to see how strategic thinking contributed to developments, taking them out of the realm either of the inevitable or of the random and unpredictable. Let's see how strategic modeling might have altered the direction the world took. Ancient Greece provides a good starting place.

SPARTA'S GALLOPING DECLINE

Sparta, you will recall, won the Peloponnesian War (431–404 B.C.), defeating Athens and emerging as the leading power in Greece and perhaps the world. Yet just thirty-three years later, Sparta was handily defeated by Thebes at the battle of Leuctra. So, in the span of a generation and a half, Sparta went from victory in the age's equivalent of a world war to a defeat from which it never recovered. How could Sparta go from the pinnacle of glory—the United States of its time, the hegemon, the greatest power on earth—to the nadir of defeat—the vanishing Austro-Hungarian Empire of its day—in a mere third of a century? The answer: They loved their horses more than their country.

Pythagoras died about three generations before Sparta defeated Athens. I mention this to call attention to the fact that the essentials of basic mathematics, especially geometry (but not probability), were readily available to educated Spartans. Their system of government emphasized education, although military prowess was a much greater focus than what we today might call book learning. Still, the Spartans could have assembled a team of mathematicians or political consultants to work out the dangers inherent in the course they followed between their great victory in 404 and their decisive and disastrous defeat by Thebes at Leuctra three decades later. Had they done so, they would have seen that their military success put their state at risk because it changed who got to vote and therefore who got to govern. As we know, voting rules can fundamentally change the very direction of politics. They did for Sparta.

To understand what happened we need to take a brief look at how Sparta was governed. Theirs was a strange and complicated form of government. Citizens, known collectively as Spartiates, were a small part of the population. By 418 B.C. the male Spartiate population fell to around 3,600 from its peak at 9,000. This was out of a total population in Sparta (including a vast majority of slaves) of approximately 225,000. After the defeat at Leuctra in 371 B.C., the Spartiates consisted of fewer than a thousand men, and it kept dropping after that. The number of people who ran the show was plummeting, for reasons directly linked to their victory in the Peloponnesian War. As we saw in Chapter 3, change begets change.

The male Spartiates elected their leaders by shouting loudest for the most desired candidates. How strong the shouts were for different candidates was determined by judges behind a curtain (or in a nearby cabin, unable to see, but able to hear the assembled citizenry) so that they did not know who voted for whom. In this way, the Spartans chose the two people who would simultaneously rule as kings (I said it was a strange and complicated form of government). They likewise chose the Gerousia (a select group of men over sixty who served for the remainder of their lives once elected), and the Ephors, who were elected to a one-year term.

The kings were in charge of military matters and national security. The Gerousia—Sparta's senior-citizen leaders—set the legislative agenda, while the Ephors had financial, judicial, and administrative power. They even had the authority to overrule the kings, while the Gerousia could veto decisions by the assembly of Spartiates. Under Sparta's system of checks and balances, Ephors could trump the kings and the Gerousia could trump the Ephors. That made it hard for any one of these elected groups to assert full control over Sparta's government.

Male Spartiates had the privilege of serving in the army, defending Sparta against its enemies. This was the driving force behind Spartiate life and the defining principle that reflected what Sparta stood for above all else. Spartan citizens were meant to be devoted to their city-state and to be better prepared than any rival to defend themselves and their society. Spartan warriors either died on the battlefield (carried home on their shields) or they returned home alive (and presumably victorious) holding their shields. Any Spartan who returned from battle without his shield was vilified forever as a coward, no matter what heroic deeds he might later perform.

In addition to military service, Spartiates were obliged to sponsor monthly banquets for their groups of fifteen, known as *syssitions*. Failure to pay one's fair share to maintain the syssition and its banquets meant losing citizenship. As the term "spartan" now betokens, the banquets were not lavish affairs. They were carefully scripted to ensure equality among all Spartiates. They were occasions for sharing with one's comrades and also providing for the impoverished masses that benefited from the leftovers.

Victory in the Peloponnesian War, however, created new ways to amass great wealth, especially among the military officers assigned to govern the lands conquered by Sparta. With the empire growing, the uneven distribution of wealth between those Spartiates who controlled colonies and those who did not steadily eroded the Spartan commitment to relative equality among the citizens. This growth in empire led quickly to two disastrous consequences.

First, the newfound wealth led to more lavish banquets. Here, like the earlier example of a failed pharmaceutical merger, the dinner menu turns out to have mattered for the future course of events. This time, however, the cost was much bigger than the failure of a lucrative business opportunity. The more upscale menu may have changed the course of history. As the price of obligatory banquets went up, many Spartiates were compelled to drop out of their syssition because they could not afford the costs. This meant that they lost their rights as citizens. So when it came time to vote for leaders, some citizens were now disenfranchised ex-citizens. They couldn't satisfy the requirements and so they lost their right to vote. The voting rules, tied as they were to providing what had become expensive banquets, shifted control over Sparta from the relatively many (a few thousand) to the few wealthiest citizens (hundreds rather than thousands).

Second, the cost of maintaining citizenship distorted careers, diminished commitments to remain in Sparta, and turned the political fabric of Spartan life upside down. Young men increasingly chose military commands outside the city-state proper rather than staying at home. They aggressively sought colonial postings because these were the path to wealth and influence. Competition for such positions further corrupted the Spartan system as these lucrative jobs were gained through patronage and cronyism instead of merit and accomplishment.

No longer was Sparta the martial—if I may, spartan—society envisioned by its founder, Lycurgus, four hundred years earlier. Instead, wealth grew in importance, whereas military prowess alone had been the dominant source of prestige before. As wealth grew among a few especially successful military officers, they pushed the cost of maintaining citizenship up, turning themselves into oligarchs. The rising price of banquets compelled more members of the Spartiate to become selfish rather than devoted to the common good. Those who were not driven by greed, or just weren't good at becoming rich, also tended to be those who could no longer pay for banquets and so couldn't maintain their rights as citizens. The consequence was that the ranks of Spartan citizens devoted to that city's founding values shrank. Those who remained became greedier and more self-centered. They needed to be if they were to survive as players on the new Spartiate stage. Greed and self-interest became the way to make a success of one's Spartan citizenship. Remember game theory's dim view of human nature? Well, here was that dim view hard at work, gradually transforming a successful society into a basket case.

What, you may well wonder, does this have to do with horses and horse racing, let alone Sparta's military defeat by Thebes? With this background information at hand, we can now answer these questions and see how game theory could have helped the poor Spartans see where they were headed, even as it predicts that self-interest will beat out the collective good just about every time.

The Greek writer Xenophon provides us with an explanation of what happened to Sparta at Leuctra. Here is what we know from him as Sparta approached its battle against Thebes. We know that the army of General Epaminondas, the leader of Thebes's military campaign, was greatly outnumbered by the Spartan army under King Cleombrotos. There were about 11,000 Spartan soldiers to only 6,000 for Thebes. The manpower advantage being with Sparta, victory should have been relatively easy, particularly because Sparta had a history of superior cavalry as well as foot soldiering. It also had a phenomenal track record of military success. What was the status of Sparta's usually exceptional cavalry as the battle approached?

Xenophon reports, regarding the contending cavalries (where Thebes had a numerical advantage):

> Theban horses were in a high state of training and efficiency, thanks to their war with the Orchomenians, and also their war with Thespiae; the Lacedaemonian [i.e., Spartan] cavalry was at its very worst just now. The horses were reared and kept by the richest citizens; but whenever the levy was called out, a trooper appeared who took the horse with any sort of arms that might be presented to him, and set off on an expedition at a moment's notice. These troopers, too, were the least able-bodied of the men—just raw recruits simply set astride their horses, and wanting in all soldierly ambition. Such was the cavalry of either antagonist.[1]

The few remaining Spartiates—the richest citizens, as Xenophon reports—withheld their best horses and their best horsemen from the risks of battle. These they thought better kept to run in races on which there would be heavy betting and lots of money to be made. Thus, Sparta's few remaining citizens, the self-centered and the greedy, chose to put their worst horses and least experienced horsemen in battle and to keep the crème de la crème for themselves. They sacrificed their city-state to preserve their personal prospects at the races. As I said earlier, the Spartans apparently loved their horses more than their country—a telling symptom of the sickness that was draining the life of the famed city-state.

Looking at Sparta's decline from this particular angle, I constructed a little data set to put into the forecasting program. It shows that the Gerousia and kings were committed to protecting Spartan security from the outset even if it meant personal sacrifices. I assumed that the Ephors started out favoring making money, as indeed it appears they did. The program shows, however, that they would quickly sacrifice their personal wealth (for instance, their horses) to protect Sparta. But what of the colonial commanders and the richest Spartiates, who remained in Sparta? They were the core of the army and the most important people deciding whether to provide their very best for the Spartan cause or for themselves (in horse terms, to use them for the cavalry or to use them for the races). The model shows them to be utterly impervious to pressure from their government, their kings, the Ephors, and the Gerousia. All the government's checks and balances, all the history of Spartan devotion to the common good, all the threats of the moment couldn't convince these citizens to do what was best for their country.

It seems that the society's newfound wealth, and the shift in people's values produced by that wealth, changed their behavior. Just as in the earlier discussion of Leopold in the Congo and in Belgium, Sparta's changing conditions led to changed behavior that in turn changed the course of Sparta's future. Had any smart Spartan looked at the data enough in advance, perhaps they would have seen the threat of collapse that their new "game" exposed them to, and maybe, just maybe, they would have thought more about the long term.

So it was that Sparta fell quickly from monumental power to weak and vulnerable backwater. There was nothing to be done to save Sparta according to the game model. The victory against Athens sowed the very seeds of Sparta's destruction.

Can we see here a larger lesson to be learned about the risky game that empire expansion might generally entail? Might U.S. efforts to spread democracy, to overthrow "rogue" regimes, come back to bite us, creating the sort of greedy, self-centered egoism at the top that brought down Sparta, or might these efforts rein in the worst abuses suffered by hapless souls elsewhere at the hands of their own greedy, grasping governments? These are questions worth pondering. History certainly has much to teach us.

Sparta's defeat revolutionized thinking in the Greek world and made possible the resurgence of Athens. The Athenians, with a more democratic—by the standards of the day—and therefore more accountable administration, were able to adjust to their earlier defeat. They could make changes that allowed them to bide their time, rebuild their strength, and reclaim

Greek leadership when Sparta faltered. Sparta's increasingly concentrated oligarchy left it with little to fall back on in the face of its unprecedented defeat. Maybe, then, we are all fortunate that Spartans loved their horses so much. Had they not, the early Greek experiment with democracy might have failed miserably, and this most beneficial form of government would have died so soon that no one would have thought to resurrect it a couple of millennia later.

Sparta's love of horses reminds me that it is time to get back in the saddle again. You know—where a friend is a friend. I have in mind the friendship that Luís de Santangel—you'll find out about him soon enough—owed to Ferdinand and Isabella of Spain and the consequences of that friendship for Christopher Columbus and even for those of you reading this book anywhere in the Western Hemisphere, the so-called new world. Santangel is the unsung hero of our next game. He reminds me of the French bankers we met in Game Theory 102. They, like Santangel, understood that demanding too much leads to nothing. Just as the French bankers put a merger together by agreeing to let the German executives remain in Heidelberg, Luís de Santangel figured out how to merge the interests of Spain's monarchs with those of Columbus, to the benefit of both.

So let's leap ahead now to the end of the fifteenth century. That time, the age of discovery, created another, new form of challenge to the dominant political order of the day. Whereas Sparta suffered from becoming rich, Spain suffered in comparison to its great-power neighbors like Portugal and the Catholic Church because it was poor. With Santangel's help, Columbus would change all of that, at least for a century or so, but not before he would have to swallow some pretty bitter pills in order to make a deal.

WHY SPAIN "DISCOVERED" AMERICA

In fourteen hundred and ninety-two Columbus sailed the ocean blue. I know, you already know that. It is curious, though, that most of us know very little about why Columbus, an Italian navigator recently employed by the Portuguese court, sailed under the Spanish flag. The story behind Ferdinand and Isabella's decision to back Columbus's journey rests on the thinnest of distinctions between rejection and acceptance. Surely had it not been for Columbus, someone else would have found the "new world," but then the course of European and American history would have been radically different. There would have been no Spanish empire, no Spanish armada for the British to defeat, probably no Sir Walter Raleigh, no Monroe Doctrine, no Juan or Evita Perón, and who knows what else. (If you will allow me a little personal view on Columbus's importance, probably without him there wouldn't even be a Bueno de Mesquita family anymore. They were pretty prominent back then, operating within the Columbus family's fiefdom of Jamaica as—of all things—pirates of the Caribbean.[2])

Columbus first put forth his proposal—to find a westward passage to Asia—to the Portuguese crown, then the world's greatest sea power. He would sail for Japan by going west from the Canary Islands. According to his reckoning, the distance to Asia was about 2,400 nautical miles. He did not think that a significant land mass was in the way of the passage, although he did expect to encounter some unknown islands. Columbus understood that there was a real risk that there would be no opportunity to get fresh water and food for his crew once they left the Canary Islands, but he did not view this as a severe problem. He believed his ships could carry enough food and fresh water for such a journey. He felt that he would have no trouble reaching his destination or returning safely. Columbus asked Prince Juan II of Portugal to fund his project—and was turned down flat.

Many factors worked against Columbus's effort to sell his ideas to Portugal. Thanks to the vision of Prince Henry the Navigator, they already had lucrative trade routes and colonial expansion along the North African coast and as far away as the Azores, about 900 miles out in the deep water of the Atlantic Ocean. Additionally, at about the same time that Columbus made his offer, Bartholomew Dias discovered the Cape of Good Hope at the tip of Africa, and therefore, implicitly, the passage up the eastern side of Africa and on to Asia. Dias was already under commission by the Portuguese government and had, by the late 1480s, discovered critical features of an eastward path to the Indies. The sea route he found offered ample opportunities for resupplying ships at coastal stations along the eastern shores of Africa. And finally, Portuguese scientists disagreed with Columbus's estimate that the journey involved only 2,400 miles. They thought that the distance between Por-

tugal and Japan going west was not much different from the distance
between these two countries going east around the tip of Africa. They
believed the distance from the Canary Islands to Japan was about
10,000 miles (it is actually about 10,600 miles), plus the additional
miles from Lisbon to the Canaries. This difference in estimates of the
distance was crucial. From the Portuguese perspective, the probability
that ships could reach Asia by sailing west was almost zero. No ship of
the day was capable of a ten-thousand-mile journey without stopping in
ports along the way for food and water. Quite simply, the Portuguese
government believed that such a journey was doomed. There just wasn't
much in Columbus's plan for the Portuguese.

Disappointed, Columbus sought support elsewhere. His brother, Bartholomew, tried to entice the kings of France and England, but they were
tied up in domestic political problems at the time and showed no interest.
Columbus approached the Spanish government in 1486. As with his
brother's efforts in France and England, Columbus found little to encourage him in Spain. The Spanish monarchy was too busy to pay much attention to Columbus. There was trouble with their Moorish neighbors and
with the pope, especially since the Spaniards had backed the schismatic
pope in Avignon.

Columbus, having no place else to turn, stayed on in Spain for the next
six years, repeatedly being put off, awaiting the findings of governmental
commissions studying his proposal. He was told that a firm decision could
not be made until Spain resolved some of its internal problems, most notably its war with the Muslim-dominated Spanish state of Granada. That
struggle finally came to an end in 1492. In fact, 1492 marked a major
turning point for Spain. The defeat of Granada in January meant that all
of the important kingdoms of Spain were now united under Ferdinand
and Isabella. At last they were ready to turn to the proposal made by
Columbus. His time had come, but the circumstances were not so much
in his favor, and he knew it.

The Talavera Commission had reported to Isabella in 1490 that
Columbus's plan was weak and advised against backing him. Like the Portuguese, members of Spain's Talavera Commission felt that Columbus
greatly underestimated the distance across the ocean between Spain and
Japan. Columbus was able to point to evidence to support his estimate of
the distance. He noted that bodies and unknown trees occasionally

washed ashore. These, he pointed out, had decayed by amounts that were consistent with his estimate of the sailing distance. Of course he did not know—how could he?—that he was about right regarding the distance to the next really big landmass, but that landmass was not Asia, it was the Americas.

Columbus grew weary of waiting and made a take-it-or-leave-it offer. Given the negative reports on his prospects of success, and given that he was demanding payment up front, Columbus found himself once again facing rejection. Ferdinand, in particular, preferred leaving Columbus's proposal to taking it. And so Columbus packed his bags and headed out of town. But this time, thanks to the intervention of Luís de Santangel, the course of Spanish-American history was assured.

Santangel, keeper of the Spanish Privy Purse, was roughly the Spanish equivalent of today's U.S. secretary of the treasury. He saw the potential merits behind Columbus's plan, and—most important—he saw a way to justify the risks by diminishing the costs. He called on the queen the very day that Columbus left Santa Fe, urging her to meet Columbus's terms. Santangel feared (apparently mistakenly) that Columbus would sell his plan to one of Spain's competitors.

Eventually he persuaded Isabella, who was more sympathetic than Ferdinand, to support the proposed journey provided that he, Santangel, would raise the money. It all came down to a matter of price. Columbus had begun by insisting that the Spanish monarchs pay the costs of the expedition in advance, including his compensation. The Spanish crown was unwilling to meet those terms. Later he reduced his price, asking only for the cost of three ships and their provisions and crew up front. He agreed to a 10 percent commission for himself and his heirs drawn from any wealth generated by his discoveries. That meant a smaller upfront cost, and it meant shifting most of the risk to Columbus and his sailors.

Of course, if Columbus succeeded, the value to Spain would be enormous: Spain would dominate a lucrative trade route and would become— as it did—a major economic force in Europe and the world. Under existing circumstances, Spain had no access to the Indies at all. They did not know of the eastward route around the tip of Africa, because the Portuguese kept their navigational charts secret. The Italians, Arabs, and others garnered the benefits from the overland caravan trade. But failure had its risks too. If the monarchs had paid out of Spain's budget, they would have faced a real prospect of rebellion by Spain's aristocrats. These very people had tried to overthrow Isabella's father just a quarter of a century earlier, so theirs was no idle threat. That is why Santangel's decision to raise the money privately was so crucial to Ferdinand's change of view. He had removed the threat of rebellion by the Spanish aristocrats, who would have resented being taxed to pay for a scheme with little chance of success. Whatever the modern take on Columbus, we surely owe Luís de Santangel three cheers for a job well done.

The Spanish decision is readily modeled with just four players: Ferdinand, Isabella, Santangel, and Columbus. As is fairly typical of a negotiated contract or business acquisition, price is a big issue. Columbus wants the most he can get. Ferdinand wants to pay nothing now; he wants it all on spec. Isabella is less negative than Ferdinand, while Santangel is supportive of the proposed exploration, but only at a price he can raise. Columbus cares most, of course, followed by Santangel and then Isabella. Ferdinand is least focused on Columbus's proposal, being more concerned with mending fences with the pope in Rome and managing his newly unified country. As king, Ferdinand has the most clout, but Isabella was no slouch and neither was Santangel. Columbus probably had relatively little ability to persuade anyone. After all, he had failed in his efforts for six years. Here, then, is the data set I constructed:

Stakeholder	Influence	Price	Salience
Ferdinand	100	0	40
Isabella	70	25	60
Santangel	60	55	75
Columbus	20	100	90

The initial prediction is between 25 (the weighted median voter) and 37 (the weighted mean voter). You can calculate those values yourself from the table. The eventual prediction after the model simulates the bargaining process and its dynamics is for an agreement to be reached at whatever price Luís de Santangel was prepared to offer as long as he was prepared to pay more than Isabella. If Santangel had offered less than Isabella's price, no deal would have been reached and world history would be quite different. Santangel was no fool. He was a skilled strategist who understood how to persuade and to whom he needed to make his pitch.

In the model's logic, Santangel first persuades Isabella and then Ferdinand to go along—at no out-of-pocket expense to themselves—and then he negotiates with Columbus. That seems pretty close to what actually happened. Of course, we don't know how much he was actually prepared to pay, only what he did pay. That's just the sort of information that my car-buying technique can ferret out, but to use that approach there have to be multiple bidders. Columbus knew—and probably Santangel, Isabella, and Ferdinand did not—that there were no other credible prospective buyers for his plan. So he probably agreed to a lower price than he could have gotten. Too bad Columbus didn't have a predictioneer available to help him negotiate a better arrangement, but the important thing is that the deal got done.

■   ■   ■

While the implications of Columbus's success are pretty apparent to us in the new world, it also had far-reaching consequences for Europe, which provides a nice little segue to our next case, World War I. You see, Ferdinand and Isabella's nephew, Philip II, was born in Valladolid and became the Spanish monarch in 1556. Philip was Spain's king during the defeat of its armada by England in 1588, which laid the groundwork for England's emergence as a great power and rival to the Spanish empire. Philip also was the Holy Roman Emperor, making him one incredibly powerful fellow. He ruled over Spain and its American empire (thanks to Columbus, Luís de Santangel, and Philip's aunt and uncle), but he also ruled over Austria, Franche-Comté, Milan, Naples, the Netherlands, and Sicily.

And so the success won by Luís de Santangel's clever maneuvering translated a generation later into binding Spain and Austria together. It also translated into sustained hostility between the French house of Burgundy and Philip's Habsburg family in Austria and elsewhere. We already know of the tension between Philip's family and England following the armada. And so, with lots of action along the way, the times of Ferdinand and Isabella had already set the foundation for "the war to end all wars" in 1914.

Let's have a look at the aftermath of the assassination of the Archduke Franz Ferdinand, heir to the Austro-Hungarian throne. We can use the history of the crisis precipitated by the Archduke's murder to ask whether the First World War (the European monarchies' last gasp) was inevitable or could have been avoided. And if it was avoidable, what needed to be done differently? We will use my new forecasting model to address these questions.

AVOIDING WORLD WAR I

The circumstances of World War I paint a truly sad picture. Wars almost always could be avoided if people just knew at the outset what the outcome would be. They can almost always make a deal before fighting starts that leaves both sides better off than is true during and after the war. This is so for the simple reason that whatever costs the combatants bear in fighting, they could avoid bearing while agreeing to the same conclusion as arises at war's end. The problem is, of course, that each side is unsure at the outset how things will turn out. They bluff about their own strength and resolve in the hope of extracting a really good deal. Much of the time that may work. Wars, especially big wars, are rare events. Sometimes, however, bluffing is tremendously costly. Rather than reveal the truth, governments sometimes fight wars they would have liked to avoid, or at least know too late that they should have avoided. One way contenders in a fight could use predictioneering is to simulate what is likely to happen before the events actually unfold so that they can avoid this very problem, maybe even saving millions of innocent lives as a result.

My new game-theory model (being applied here for the first time) shows a number of ways the First World War could have been avoided. Tens of millions lost their lives because a handful of diplomats played their cards poorly. That is the essence of Greek tragedy in modern times. Before I address how the war might have been avoided, let me provide a brief background on the circumstances that led to it.

Taking a very broad and long view, it is evident that the eighteenth and nineteenth centuries were a time in which those who adopted more democratic forms of government and more capitalist economic modes, such as the Netherlands and England early on and France later, enjoyed burgeoning wealth and influence in the world. Monarchy seemed to be in decline.

Zooming in on the latter half of the nineteenth century, we see the unfolding struggle to control Europe's destiny. Germany as we know it did not exist for most of the nineteenth century. Instead, modern-day Germany was divided into many princely states—Prussia, Saxony, Baden, Wurttemberg, and many others. Austria dominated German affairs.

All of this was to change with the rise of Otto von Bismarck as Prussia's minister-president (they certainly went in for awkward titles). Bismarck built Germany into a European power. First he united Prussia with several smaller German princely states to fight the Seven Weeks' War (1866), in which, to the surprise of most European leaders, he quickly and easily defeated Austria. This war marked the end of the post-Napoleonic Concert of Europe system of checks and balances that had been forged half a century earlier to ensure stability among the great powers of Europe (then consisting of Austria, England, France, Prussia, and Russia) and to prevent the rise of another Napoleon.

The Seven Weeks' War revealed that Austria was much weaker than its status as a European great power implied. Desperate to maintain itself among the ranks of important states, the Austrian government agreed to a merger with Hungary that resulted in the creation of the Austro-Hungarian Empire. This was the same merger deal that the Austrians had rejected just before their 1866 defeat. The creation of Austria-Hungary helped keep Austria in the running as a great power, slowing but not reversing its declining political position. Just four years later, Bismarck went to war against France, defeating Napoleon III in the Franco-Prussian War of 1870–71. With France's defeat, Bismarck succeeded in unifying the remaining German princely states, creating modern-day Germany. Whereas Austria had dominated the pre-1866 concept of Germany, it was now excluded, not to be reunited with the rest of Germany until Adolf Hitler—an Austrian by birth—rose to power about sixty years later. By 1871, Bismarck had established Germany as the rising power of Europe and helped France to join Austria (now Austria-Hungary) as a state in decline. This set the stage for the First World War.

Revolutions against monarchy and oligarchy were bubbling up everywhere. There was revolt in Russia in 1905, in Mexico in 1910, and in China in 1911. From the Austro-Hungarian point of view, the most threatening emerging nationalist challenge to monarchy came from the Balkans. There the Austro-Hungarians saw in the experience of the Ottoman Empire the foreshadowing of their own demise. The kingdom of Serbia tripled its territory as a result of the Balkan Wars (1912–13), becoming a magnet for Serbian nationalists. They wanted all of Serbia out of the Austro-Hungarian Empire. Those tensions burst to the surface on June 28, 1914, with the assassination in Sarajevo, today the capital of Bosnia and Herzegovina, of the prospective heir to the Austro-Hungarian throne, Archduke Franz Ferdinand.

The assassination prompted Austria-Hungary's government to issue an ultimatum to the Serbian government: give up your sovereignty, or it's war. The Serbs were not without friends, and of course they were reluctant to give up their hard-earned independence. It seems that the Austrians were counting on this. The diplomatic records of the day, now open to us, reveal that they chose to make demands that they were confident could not be accepted. Apparently, the Austro-Hungarian leaders wanted a little reputation-building war with Serbia.

Europe's great powers chose sides in the dispute. Russia sided with the Serbs. Under the terms of the Triple Entente—an alliance between Russia, France, and England—France and England also chose to side with the Serbs. The Russian decision triggered a response from Germany, by then Austria-Hungary's ally. Under the terms of their alliance with Austria, Germany backed Austria-Hungary. Their broader alliance ties meant a high likelihood of additional support from Romania, Turkey, and especially Italy. The Dual Alliance of Austria-Hungary and Germany had been expanded in 1882 to include the new European power of Italy, with the expanded alliance referred to as the Triple Alliance.

Fearing an aggressive move by Germany in defense of Austria-Hungary, the Russians mobilized. They intended, in game-theory terms, to send a signal that they were committed to Serbia's defense. This prompted a similar mobilization by Germany. Much as in the prisoner's dilemma game we discussed earlier, each side could see that conciliation was better than war, but they also could see that trusting their adversary to pursue a settlement was risky. And so they found the result that follows from the logic of the prisoner's dilemma: they fought instead of settling. In a few short weeks, the conflict over Serbia escalated to involve all of the great powers on the Continent. World War I had begun. The little reputation-building war between Austria-Hungary and Serbia was not to be.

Shortly, I will apply my model to inquire what might have happened if the Triple Entente of Britain, France, and Russia had been more able in

1914, or, for that matter, if the Dual Alliance of Austria-Hungary and Germany had been more skillful. First, however, let's pretend that there was a little army for hire of people with good math skills pounding out my calculations in 1914. What would they have predicted with no advantages of hindsight?

To address the 1914 crisis—not the fighting of the war, mind you, but the diplomatic run-up to war—I constructed inputs for the computer program that measure the degree to which each of the European countries, plus important non-Europeans including the United States and Japan, favored Serbia's or Austria-Hungary's foreign policy in 1914. I estimate salience based on a mix of expert judgments and geographic proximity to the Austro-Serbian crisis. Potential influence is based on a standard measure of "national power" collected for every country in the world for every year from 1816 to roughly the present by the academic enterprise, introduced earlier, called the Correlates of War Project. Of course, I use the estimates for 1914. Since I am examining this case using my latest model, I include the additional variable it requires. This variable measures the extent to which each player is resolute in the position it has taken even if that means a breakdown in negotiations or, conversely, is sufficiently eager for an agreement that it will show considerable flexibility in its approach to negotiations. Put in terms of our earlier discussion of health care, this new model includes an input that calibrates how much a player's bargaining style looks like Bill Clinton's (100 on this variable's scale) or like Hillary Clinton's (0 on this variable's scale) back in the early 1990s. This "commitment" variable's values are based on my reading of the historical record in the run-up to World War I. Any interested decision maker in 1914 would have had access to the information used here. And if they had my equations, they could have done the exact analysis I report.

Austria-Hungary, Germany, Romania, and Italy start off at position 100, indicating a full endorsement of Austria's position against Serbia following Franz Ferdinand's assassination. Serbia and Greece start off at a position of 0 on the issue scale, indicating their total opposition to Austro-Hungarian demands for Serbia to surrender its sovereignty. The rest of the European states fall approximately between 33 and 45 on the scale, suggesting that they tilted toward Serbia and against Austria but not decisively.

As seen in figure 9.1, a palace full of bearded mathematicians crunching away on the numbers in 1914 would have anticipated war. They also

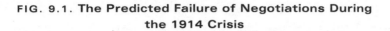

FIG. 9.1. The Predicted Failure of Negotiations During
the 1914 Crisis

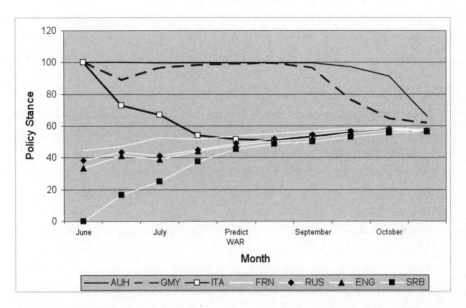

would have realized that war could be avoided if they just crunched the numbers long enough, reflecting a prolonged diplomatic effort instead of a rush to war. The figure shows that the model anticipates war sometime in August 1914. This is the stage at which the model's logic says diplomatic efforts to resolve the dispute without resorting to the use of force would have ended.

You see that the model is constantly calibrating the expected benefits from continuing to negotiate, weighing them against the model's estimate of the expected costs from continuing to pursue diplomacy. Eventually, in the absence of an agreement, the players conclude that the prospects or value of a future agreement just isn't worth the effort. In essence, the model's algorithm makes a judgment about the value the players attach to extracting a concession tomorrow compared to extracting the same concession today. Getting some benefit sooner is always worth more than getting it later. In this instance, the period that corresponds approximately to early to mid-August happens to be the time the model says the game would end because too little progress was being made in closing the gap between Austrian demands and Serbian concessions. So the game predicts its own end in August. At that point there is no agreement between the main antagonists, and so, according to the model, a new game starts, with generals taking over from the diplomats.

Up to this point the Austrians (supported by their German allies) have persisted in demanding enforcement of the Austrian ultimatum. Meanwhile the Serbian government has gone a good distance toward meeting many of Austria's demands. Still, Serbia shows no willingness to accept the Austrian ultimatum, exactly as the Austrians hoped. Instead, Serbia adopts a moderately conciliatory posture that is consistent with the concessions pushed by the British, French, and Russians. The latter three, according to the simulated crisis, were believed to be strongly committed to finding a settlement. Because of that, neither in reality nor in the simulation did the Austrians and Germans think their foes in the Triple Entente were likely to go to war on Serbia's behalf.

What were the Italians, members of the Triple Alliance, up to during the crisis? In reality, they indicated on July 28, 1914, that they could not support the Austro-Hungarian ultimatum, delivered just five days earlier. With war imminent, the Italians declared themselves neutral. They resigned from the Triple Alliance on the grounds that Austria-Hungary was launching an aggressive, not a defensive, war.

In the model's assessment, the Italians start out at the same position as the Germans and Austrians, befitting their membership in the Triple Alliance. As can be seen in the figure, by mid-July in model time the Italians break from the Triple Alliance and become neutral. They adopt a position hovering around 50 on the issue scale. So the model sees the Italians moving a week or two earlier than they actually did, but nevertheless it foresees their shift to a neutral position. In the model's logic, and in reality, the Italians were not committed to standing by their Austrian and German allies once they recognized that the unfolding events were going to produce something vastly larger than a small Austrian-Serbian war.

The model indicates that Austria-Hungary and Germany expected war over Serbia for the first few weeks of the crisis, while Serbia shared that expectation. Austria actually declared war on Serbia at the end of July while professing to have no quarrel with others. By the start of August the model anticipates that Serbia was relegated to a relatively minor role as events ran ahead of Austria-Hungary and Germany's ability to cope with them. Germany in fact declared war on Russia on August 1, and the war anticipated by the model's logic actually began.

Was war inevitable? Emphatically the answer must be *no*! First, we can see in the figure that had the contending parties sustained negotiations one or two months longer, postponing the decision to go to war, the Germans would have better comprehended the dangerous big picture. They would have (according to the model's predictive logic) broken ranks with the Austrians and come to agreement with the British and their allies. The Austrians would have gotten the majority of what they wanted under the agreement that the model indicates could have been reached in September or October 1914. Of course, this agreement needed the diplomats to remain in charge instead of turning choices over to the generals. The deal that could have been struck would not have included the surrender of Serbian sovereignty. But all that is beside the point since, alas, the game simulation suggests correctly that the diplomats would not have continued negotiating through the early fall. The diplomatic game ends before September and a new game, war, begins.

The beauty of a model is the freedom it gives us to ask lots of what-if questions. We can replay the World War I diplomacy game, just as I did in an earlier chapter for a litigation client, while changing how players present themselves. That way we can see if this or that player could have approached the game better, producing a happier result from its point of view.

Let's replay the 1914 crisis, this time making the British diplomats more skillful than they actually were but no more skillful than they could have been. I am going to let them look inside the model's approximation of what was going on in the heads of the German and Austrian decision makers. In this way I am going to pretend that they had a little army of mathematicians doing the calculations my computer does for me. This will make it easier for the British to be more thoughtful and decisive, instead of as wishy-washy as they were.

The historian Niall Ferguson has argued that a big factor leading to war in 1914 was that the Austrians and Germans were uncertain of British intentions and that this uncertainty was caused by the British.[3] Britain may have done well for a long time by muddling through, but that was not much of a strategy in 1914. Did the British really intend to defend Serbia, or were they bluffing? Certainly nothing they said or did at that time was sufficient to convince the powers of the Dual Alliance that defending Serbia was really important to Britain. This was an important failing on their part, and it deserves further exploration.

Remember that when we looked at a lawsuit I worked on, we examined the consequences that followed when I advised my client to bluff having a stronger commitment to their bargaining position than in fact they had. Such a bluff can be risky and costly. If the other side believes—correctly— that a tough posture is just posturing and not the real thing, then they will call the bluff. In the lawsuit, that would have raised the odds of a costly outcome. The client would have faced severe felony charges. They might have been exonerated in court, but trials are, as we've seen, always risky business. Without bluffing, they were going to face those charges anyway, so bluffing looked (and proved) attractive.

Think how much costlier and riskier bluffing could have been for Britain in the summer of 1914 than it was for my client in the lawsuit. With hindsight we know that the guns of August were not stilled for more than four years. At the end of the war, the United States—not Britain, not France, not Germany, and not Russia—would be the greatest power in the world. At the end of the war, Austria-Hungary would not even exist. But when decisions had to be made, no one knew any of that. They had to think about what their circumstances would look like if they showed eagerness to compromise or if they showed real resolve to stick to their guns, so to speak. The British looked for compromise, and disaster followed. What does the model say would have happened had they bluffed being resolved to defend Serbia, and how could they have conveyed such resoluteness?

The British were in an odd position. It seems that even they were unsure how resolved they were. They were uncertain not only about others, but apparently even about themselves, about what they should or would do. That, presumably, is why the Austrians and the Germans did not read British diplomacy as signaling real commitment to defend Serbia. We also know that when the Russians—believing they were facing an imminent attack—mobilized, this prompted the Germans to do the same, and war began. The Russian mobilization certainly showed their commitment, but it did nothing to improve the prospects of a negotiated settlement. Their mobilization was a very costly "costly signal." Would British mobilization have had the same dangerous consequences, or could it have broken the impasse?

The data going into the model treat Britain as highly committed to finding terms that all sides can accept. They were assigned a value of 90 out of a possible 100 on "flexibility/commitment," indicating they really wanted to negotiate and were prepared to live with a major compromise to avoid war. I have repeated my earlier simulation of the crisis, but with one change. I shifted Britain's commitment to compromise from 90 to 50. A value of 50 signals a balanced approach. A value of 50 means the player actively pursues a settlement but is sufficiently resolved that it will not make a deal very far from its desired outcome. By placing Britain at 50 I am, in essence, trying to test Niall Ferguson's insight (and that of other historians too) that the wishy-washy British message contributed to the war. I am simulating an approach that the British leaders probably would have seen as a bluff intended to shake up the situation and promote a war-avoiding deal.

What concrete actions might the British have taken to send the message, "We are serious about defending Serbia's sovereignty"? I am not a military expert, so my speculation will be just that. I am sure a military specialist or historian of British policy in the run-up to World War I would find countless other ways for the British to send the right message. Here is one:

Britain was the world's greatest sea power (although the Germans were certainly challenging that claim at the time). They could have filled several of their navy's ships with a few thousand British troops to be transported to the Adriatic, taking them just a short distance from Serbia. Maybe they could have sent some other ships into the Bosporus, roughly flanking landlocked Serbia from either side. This would have served several potentially advantageous purposes. It is very much, in game-theory lingo, a costly signal. Talk is cheap, but sending a fleet into a prospective combat zone is putting your money where your mouth is.

The Germans and Austrians probably would have taken more seriously the prospect that Britain meant business. As we will see, the model indicates just that. Additionally, a naval mobilization of this sort has none of the grave risks associated with the Russian mobilization of ground forces. Russia could move troops quickly to and across the German frontier. Understandably, that made the Germans more than a little jittery. British ships filled with soldiers would have taken a long time to get into position. Finally, the ships would not have been directly in the path on which initial fighting was expected. Thus the shiploads of troops would very much have been a signal of what was to come without precipitating immediate military action. In reality, British ships under French command headed for the Adriatic a few days after war had been declared: too little too late.

When I simulated the prewar 1914 crisis with the British at 90 as their "flexibility" variable, the model indicated that the Austrians and Germans were as uncertain about Britain's true intentions as they could be. But when I place Britain's score on this one factor at 50, the model shows that the Germans and the Austrians are convinced that the British will fight. More tellingly, Austria-Hungary's and Britain's pattern of interactions change. When Britain shows little resolve, Austria-Hungary anticipates coercing the British into accepting their position. When Britain shows stronger, but not extreme, resolve (50), the Austrians seek a negotiated compromise with the United Kingdom even as they perceive that if Britain is allowed to move first it will mean war. Not, mind you, the little war that Austria sought with Serbia, but the big war that nobody wanted.

Have a look at figure 9.2. Here we see the scenario in which Britain shows more resolve (50 instead of 90). By simulating a tougher British signal, we uncover the sort of intelligence about an adversary's thinking that would be a lifetime coup for a real-world spy. We discover that with British ships heading for the Adriatic shortly after the start of the crisis, well before war is declared, Austria-Hungary, Germany, and the U.K. perceive the possibility of settling their differences quickly. The Austrians and Germans saw no such prospect (or reason for it) in the simulation of the actual situation. But with the U.K. showing more resolve, the strategic environment is greatly altered. The Austrians and Germans believe that they can and should make a deal with the U.K. in a matter of days after the start of the crisis.

The model says that the Austrians and Germans recognize that they should give up their demand for utter Serbian capitulation. They see an opportunity to persuade the Triple Entente to agree that Austria-Hungary should have real, but not controlling, influence over Serbia's foreign policy.

Of course, change begets change. The members of the Triple Entente don't immediately embrace the new offer that (according to the model's logic) would have been put on the table if the British had sent navy ships to the Adriatic right away. While not caving in to this simulated proposal, the Triple Entente's diplomats certainly sit up and take notice. The diplomats remain in charge, keeping the generals on the sidelines. While the negotiators hold out for a few weeks, thinking they can extract more concessions from Austria-Hungary, by early August they realize (in the model's logic—remember, none of this was actually done in 1914) that

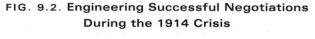

FIG. 9.2. Engineering Successful Negotiations
During the 1914 Crisis

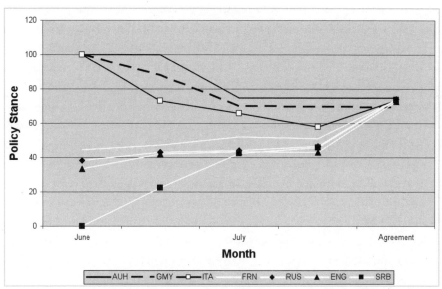

the Austrians and Germans are highly reluctant to give more. So they take a deal located at about 75 on the 100-point issue scale. That would have given the Austrians more than was put on the table in the real 1914 crisis, but a lot less than the surrender of Serbian sovereignty. It would have more or less split the difference between the British position and the Austro-Hungarian demand. The simulation under these conditions shows that the French and the Russians would have quickly acceded to this compromise. The war to end all wars would have been avoided.

Had there but been a thousand mathematicians crunching numbers in London in June 1914, we might not need to ask the next and final question of this chapter: Could World War II have been prevented by the judicious use of a predictioneer's skills?

NO MORE MISTER HITLER?

The rise to power of Adolf Hitler is a strange and horrendous tale that could have been nipped in the bud sometime between November 1932 and March 1933, if not sooner. It is a tale worth heeding. Whatever else can be said of Hitler, it must be admitted that he was honest and open about his intentions. Not only did his biography, *Mein Kampf,* published in 1925, announce his ravings to the world; so did he, in campaign speech after campaign speech in 1932. Having failed to come to power in 1923 through an attempted coup d'état (the Beer Hall Putsch, as it is known) in which both German police and Nazi insurgents were killed (and Goering seriously wounded), Hitler was now determined to rise to dictatorial control through legal means. The ballot box would replace the bullet for the moment.

In one campaign stop after another—several per day, as Hitler was the first German politician to take advantage of air travel to cover vast amounts of ground—Hitler declared his intention to ban political parties and suspend the Reichstag, Germany's parliament, if he came to power. Now, when a campaigner promises peace and prosperity, motherhood and apple pie, we don't really learn anything about what they plan to do. Being for hope or change or a thousand points of light says nothing. When a politician promises to overturn democracy, that's a different matter. You don't lose votes by promising peace and prosperity. I suppose being denied freedom of choice appeals to some people, but certainly not all and probably not many. So when a politician makes such outrageous declarations, we must ask: What does he have in mind? The answer must be that he means what he says. Hitler certainly did.

Of course, we have little need to pay attention to every fringe movement that makes outrageous declarations. But by 1932 the Nazi Party was no fringe movement. The 1930 German election—they ran elections with incredible frequency—gave it 107 out of 577 seats in the Reichstag. In the July 1932 election, the Nazis became the single largest party in the Reichstag, with 230 seats. By then no prudent person could treat Hitler's campaign promises lightly. Anyone listening should have understood, given the nature of what he was saying, that he meant what he said and he said what he meant. Hitler was a dictator at heart, one hundred percent.

Hitler's party lost some seats in the November 1932 election but still remained the single largest contingent in the German parliament. The election gave the Nazis 196 seats. By January 1933, Paul von Hindenburg, the beloved and elderly German war hero and president of the Weimar Republic, acquiesced under pressure to make Hitler chancellor. Now the door was wide open to his dictatorial ambitions. In early March 1933 another election was held, almost immediately after the Reichstag fire on March 3. Hitler was quick to blame the Communists for the fire, using that as a pretext to ban them from the Reichstag. He wanted all Communist leaders executed that night—no more Mister Nice Guy—but Hindenburg refused to go along.

The March election was a mixed success for the Nazi Party. On the plus side (from its perspective), the party increased its seat share in the Reichstag from 196 to 288. On the minus side, the Nazis failed to gain a majority. That meant Hitler still had to make deals with other parties; he was not yet completely in control. He was still vulnerable to defeat if a strong enough coalition of parties in the Reichstag joined together to oppose him. The tragedy is that they didn't.

Shortly after the March 5 election, on March 23, 1933, he negotiated his way to a two-thirds vote in the Reichstag to change the constitution to comply with the terms of the Enabling Act, a piece of legislation to make him dictator. The Enabling Act gave Hitler as chancellor essentially all of the constitutional authority also granted to the Reichstag so that he no longer needed legislative approval for policy changes he wished to put in place. The Enabling Act made Hitler Germany's dictator and ended the need for future elections. He was well on the way to doing everything he had promised to do during the previous election campaigns.

Once the Enabling Act was approved, probably nothing short of a military uprising or foreign military intervention could have stopped Hitler on his destructive course. What about between November and March? As I said, Hitler's intentions were no secret. Could he have been thwarted before the Reichstag fire resulted in the one hundred Communist Party members in parliament being banned, making it much easier for Hitler to put together a two-thirds vote? Hitler was stuck operating within the legal system, at least more or less, before the Enabling Act's passage. It was far from a sure thing that he could muster a two-thirds vote. The key to his success—or failure—was the Catholic Center Party (BVP).

Allow me for a moment to frame the game as it was set up. There were four principal parties in the Reichstag at this time: the Nazis, the Catholic Center Party, the Social Democrats, and the Communists. The Social Democrats and the Communists generally opposed Hitler, and would resist his Enabling Act (the Communists, of course, irrelevant in the actual voting on account of being barred because of their supposed role in the fire). The Catholic party, however, was divided over whether to support the Enabling Act. Their leader, Ludwig Kaas, was no fan of Hitler's. Nevertheless, Kaas, a priest, negotiated with Hitler for assurances that Catholic interests—inside and outside government—would be protected if Hitler were given their support. Kaas may also have sought assurances of a concordat with the Vatican. Hitler agreed to his terms, understanding the temporary need for compromise.

Hitler's deal with the Catholic party was the crucial decision that gave the Nazis the two-thirds majority they needed for the Enabling Act. With the Communists barred, the Social Democrats were the only party to vote against the act, and of course Hitler won the game as a result. But it remains the case that without the Catholic Center Party, the Enabling Act would have been defeated, Hitler would not have been made dictator, and, who knows, perhaps the course of world history would have been completely altered.

Beating Hitler was no easy task. We must admit that he played even the parliamentary game the best out of all the parties. His opponents either underestimated him (with fatal consequences) or were simply incapable of outplaying him. That, however, is not to say that there weren't winning moves open to them.

What or who could have stopped the Catholic Center Party from going along with Hitler? How could they have made someone—anyone outside the Nazi Party—Germany's new, democratic boss? Can you see a possible strategy?

I applied my forecasting model to this question, constructing a data set in which the stakeholders are the political parties represented in the Reichstag plus Hindenburg. Their power is proportional to the number of seats they controlled, except, of course, Hindenburg, who had no seats in the Reichstag. His personal prestige and popularity, however, gave him great weight, even more than Hitler and his Nazis. Remember, Hitler could not execute the Communist leadership without Hindenburg's (withheld) approval. I accordingly assigned Hindenburg 67 percent more clout than the Nazis. Positions on the Enabling Act are and were well known at the time. The Communists and the Social Democrats were utterly opposed, the Nazis utterly in favor; Hindenburg leaned favorably toward the Act and the

Catholic Center Party did too, but ever so slightly. The rest were committed to the Act.

Right after the November 1932 election, while Hitler was angling to become chancellor, the Social Democrats and the Communists could (according to my model) have struck a deal with the Catholic Center Party, depriving Hitler of the two-thirds majority he needed. To do so, however, they would have had to move meaningfully in the direction of the Catholic Center Party's policy desires, perhaps so much so that a member of the Catholic Center Party would have become the chancellor instead of Hitler. That is, the Social Democrats and the Communists needed to provide at least as good an assurance of Catholic interests as Hitler astutely provided months later. The model indicates that they did not believe they had this opportunity. They did not think that the Catholic Center leadership would listen to them or make a deal with them, and so, fearing rejection, they didn't really try (or at least they didn't try hard enough). The model says they were wrong. Too bad we can't go back in time to test the waters and see if the deal could have been made.

We do know that Ludwig Kaas, the Center Party's leader, had in the 1920s developed good relations with the Social Democrats' then leader, Friedrich Ebert. It is difficult to imagine that under the circumstances Kaas would not have responded to the Social Democrats, especially if they were prepared to support him as the next chancellor (let's remember game theory's view of human nature and what it means for individuals seeking to obtain or maintain power and influence). The model says Kaas would have reached a deal with the Social Democrats and the Communists. Of course, for the Communist Party, atheists that they were, bitter adversaries of both the Social Democrats and the Catholic Center Party, this would have been a bitter pill to swallow. But surely it was better than the easily foreseen fate they met after the Enabling Act was passed. Many Communists were murdered, others sent to concentration camps.

Even after the Reichstag fire, the Social Democrats still had the chance to cut a deal with the Catholics, but didn't (although whether Hitler could still have been stopped by that stage is questionable). Yes, he could have been deprived of a two-thirds majority, but just barely, once the Communists were out of the picture, and he was, as I noted earlier, playing his strategic cards effectively. In November, December, and maybe through

January, however, there is a good chance that a defeated Nazi Party would have been relegated to the dustbin of history. Had Hitler attempted a coup at that stage, it probably would have failed. The German security apparatus would have rallied behind a Catholic Center–led government. The German general staff had no love for Adolf Hitler and his brownshirts, and he was a long way from being the undisputed representative of the people. Remember, the Nazis lost seats in 1932. German popular will was not yet decisively behind Hitler or his party.

Without Hitler, the capitalist, democratic world and the communist, totalitarian world led by Joseph Stalin might have butted heads in the 1930s instead of waiting for the cold war. Perhaps even as bloody a war as World War II would have been fought, and perhaps not. Stalin's Soviet Union might have successfully defended its frontiers, but it probably had nothing like the capabilities of Hitler's rearmed Germany to wage an aggressive war beyond its borders.

All this is speculation, of course, but there seems little reason to doubt the model's accuracy in this or the other cases looked at in this chapter, given its track record over thousands of applications. And if we can replay the past accurately and find ways to improve it, as we just did with World Wars I and II, there is no reason to doubt that we can fast-forward the present and work out ways to make it turn out better. That is the whole purpose of this forecasting and engineering enterprise.

In the next chapter, we will play with some of the big issues of our time. I will use my newest model to make live predictions whose accuracy you will be able to check for yourself.

10

DARE TO BE
EMBARRASSED!

I T'S ALWAYS NEAT to think about what-ifs, to rewrite the past with
the idea that we could have worked things out for the better. But thinking about how to rewrite history is one thing; thinking about how to write a
script for the future is quite another. It's so easy to get the past right when
you already know what happened. And while examining alternative pasts is
fun and informative, we still never get to find out whether we really could
have derailed Sparta's defeat or stopped Hitler in his tracks. Ultimately,
solving a seventy- or eighty-year-old problem is fascinating, but not terribly
useful outside of what it teaches us about the gaming process. Working out
how to solve today's problems, like stopping al-Qaeda dead in its tracks—
now *that* would be useful. That's why any predictioneer worth his or her salt
must be willing to risk the embarrassment that comes from being wrong.

In this chapter we will look ahead a year or two from when I am writing
this (in April 2009 for the case of Iran-Iraq relations and June 2008 for the
Pakistan case). Here is where the rubber really meets the road. We will
look at what the United States government could do to diminish the threat
of terrorism or insurgency in Pakistan, and the likely relations to develop
between Iran and Iraq if President Obama fully withdraws U.S. forces
from Iraq or leaves fifty thousand in that country well beyond August 2010.

■ ■ ■

Back in the spring of 2008 and again in 2009 I taught an undergraduate
seminar at NYU in which twenty terrific students in each class used my

new forecasting model. This was a great opportunity for me (as well as, I hope, for my students) to find out how hard or easy it is to teach people with no prior experience how to become effective political engineers. Fortunately, my students were willing guinea pigs, and they did a great job.

The main idea behind this course, sponsored by NYU's Alexander Hamilton Center for Political Economy, was to search for solutions to pressing policy problems based only on logic and evidence. That is the Center's mission. It leaves no room for partisanship, ideology, opinion, anecdotes, or personal wishes when it comes to crafting solutions. Game-theory models, however, are a way to fulfill the mission. With that in mind, I asked my students to pick any foreign policy problem that intrigued them. They clustered themselves into groups and set to work on Pakistan, the Israeli-Palestinian dispute, global warming, nuclear proliferation, relations between Cuba and the United States, relations between Russia and the Ukraine, and many other critical policy concerns.

Each student studied a problem that he or she really cared about. They took the class knowing that they would use game theory to work out likely future developments and to write a script about how to improve the future from the perspective of any one of the players in the game. They had almost no prior experience with any of the material or models. They had limited access to experts, so they relied on the Internet and major news outlets to put their data together. I mention this to be clear that any hardworking, motivated person can replicate what they did. All this being said, my students used my new model, and I certainly reviewed their work—so any misses are the model's and mine.

Okay, let's see what they came up with, remembering that the first class began in January 2008 and had its last meeting on May 5, 2008, and that the second started in late January 2009 and ended in the first week of May 2009. Everything reported here was worked out during those months. No information has been updated or altered to take account of later developments. The students had no prior experience with my old forecasting model or my entirely different and more sophisticated new model. We met for two and a half hours each week in class. They made weekly presentations, got lots of feedback, and spent a fair amount of additional time with me in my office learning how to interpret the new model's results. They also put in lots of additional time figuring out what questions to ask and how to frame them, assembling the data, and preparing their weekly presentations and final papers. Let's have a look at what they found out.

PAKISTAN: WHERE HAVE ALL THE SOLDIERS GONE?

The group that decided to work on Pakistan in 2008 was intrigued by three policy questions. They wanted to know how willing the Pakistani government was going to be to pursue militant groups operating in and around Pakistan, including al-Qaeda, the Pakistani Taliban, and the Afghan Taliban. They also wanted to investigate whether the Pakistani government would allow U.S. military forces to use Pakistani territory to launch efforts to track down militants. Finally, they wanted to forecast the level of future U.S. foreign aid to Pakistan and whether a higher or lower amount of aid was likely to change the Pakistani leadership's approach to pursuing militants.

These are big questions that go to the heart of U.S. interests in Pakistan. While answering these questions, the students also uncovered answers to a bunch of other important and compelling issues.

By way of background, it is important to remember that when the students started their project, Pakistan was in the midst of a crisis. Benazir Bhutto, Pakistan's former prime minister, had returned from exile in late 2007 as part of a deal she negotiated with Pakistani president Pervez Musharraf. She was expected to become the next prime minister following the general elections scheduled for January 8, 2008. Instead, she was assassinated on December 27, 2007.

The elections were postponed to February 18, 2008. Musharraf's party was routed, while the parties of former prime ministers Bhutto and Nawaz Sharif, also recently returned from exile, won control of the national assembly (Pakistan's parliament). Musharraf continued as president. Mrs. Bhutto's husband, Asif Ali Zardari, became the new head of her party, the Pakistan Peoples Party (PPP), while the party of former prime minister Nawaz Sharif, the Pakistan Muslim League–Nawaz (PML-N), joined the PPP to form a coalition government.

Neither of the two victorious parties was a friend to Musharraf. He was in the awkward position of having to certify the new government, expecting that it was likely to impeach him as soon as it was certified. He had earlier fired the chief justice of Pakistan's supreme court to prevent him from ruling on the legitimacy of Musharraf's own reelection. The new government was expected to restore the chief justice and had declared its intention to depose Musharraf as soon as the PPP and PML-N came to power. It did not do so. Failing to get PPP support on this important issue, Sharif and the PML-N withdrew from the coalition on August 25, 2008. With pressure mounting from the United States to do something about the use of Pakistani territory as a base of operations for al-Qaeda and the Taliban, and with the Pakistani government itself deeply divided on how to move forward, the country was in turmoil and its future direction was extremely uncertain.

The situation seemed dire not only for Musharraf, but from the American perspective as well. For all of his limitations, Musharraf was an important ally in the war on terror. He had literally put his life on the line by siding with the United States against the Taliban government in Afghanistan after 9/11. By 2007, however, his support seemed to waver. He turned greater authority over the pursuit of militants to local tribal officials along the Afghan-Pakistani border, reducing the role of the Pakistani army. From the American perspective, this hurt the prospects for continued success against the terrorists. Musharraf argued that it would prove beneficial because the locals knew the situation on the ground infinitely better than any outsiders and they had the local clout to get things done. (My own view was that this was a move by Musharraf to extract more economic and military aid from the United States by threatening to allow the situation to deteriorate if the aid was not forthcoming. It's important to note that my students knew nothing about my personal view.) Although very much his own man and hardly perfect in American eyes, Musharraf was nevertheless the United States' best source of help in the effort to defeat al-Qaeda and the Taliban. The new government in parliament, in contrast, spoke openly of finding a way to negotiate with the groups identified by the United States government as terrorists.

So what did my students find out? Their analysis showed that the PPP would have even greater policy influence relative to Sharif's PML-N than the PPP's advantage in National Assembly seats implied. This was not particularly surprising, but then, if a model only produces the unexpected, we should be suspicious of it. Why wasn't it surprising? To start with, Sharif did not personally compete in the February election. That left his party without strong leadership in the National Assembly. When he finally decided to run in a by-election, the courts ruled that he was not eligible because of his earlier conviction on corruption charges when he was prime minister. It was, in fact, his earlier corruption case that had sent him into exile in the first place. That aspect of my students' analysis merely confirmed what any Pakistan watcher already knew. Sharif was not as popular as Bhutto, and neither was his political party as popular as hers. With her assassination the PPP gained even greater influence, riding the crest of an upsurge in sympathy for her, her political movement, and her vision for the future.

What was surprising, and distressing, was the pattern of evolving power among the national leadership that emerged from the analysis. While capturing the conventional wisdom about the relative power of the PPP and the PML-N, they also found a solid answer to a pressing question. Many Pakistan watchers speculated on whether the new leadership in the National Assembly would make a deal with hard-liners. None, as far as I know, had dared to quantify what that might mean in terms of the future distribution of political power in Pakistan and its implications for shaping policy.

The predictions my students made, based on game-theory logic and the data they amassed to seed the model, can be seen in the figures that follow. The first addresses Musharraf's potential to survive the election's outcome and the conditions that ultimately would lead to his ouster. At the time their study was done, I think it is safe to say, most people believed Musharraf was finished. A few speculated on whether the United States would somehow save him, but most thought he would be political history right after the February 2008 election. "Not so fast," said my students' results.

Figure 10.1, on the next page, tells the story of what could have been and of what was to be. If the two parties in the government, Zardari's PPP and Sharif's PML-N, had been willing to work together, then figure 10.1 shows us that Musharraf could indeed have been ousted in March or April 2008, just as the pundits expected. The government parties' combined power—the heavy solid line in figure 10.1—overtakes Musharraf in the period between March and April. That would have been the opportunity to kick him

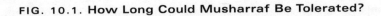

FIG. 10.1. How Long Could Musharraf Be Tolerated?

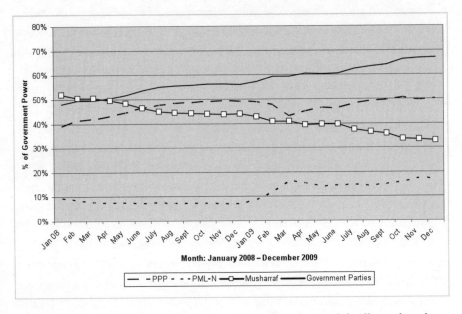

out, as expected by most Pakistan watchers. But the model tells us that the PPP and PML-N were not willing to work together at that time. The model shows that Sharif believed throughout this period that he could pressure Zardari and his party into doing just what Sharif wanted. The model also tells us that Sharif was wrong. According to the model's results, Zardari saw no reason to listen to Sharif since the PPP had substantially more clout than Sharif's PML-N. As we know now, rather than work together, Sharif threatened to withdraw his support for the government in May 2008, because Zardari was unwilling to commit to deposing Musharraf.

But figure 10.1 tells a more complete story than just that. We can also see that the model projected that Zardari's PPP (the dashed line in the figure) on its own, without help from Sharif, would surpass the declining Musharraf in power by June or July 2008. At that point, the PPP didn't need anyone's help to dump Musharraf. They had the clout to do it on their own. (We now know that they in fact did push him out in August 2008 and that Zardari assumed Musharraf's role as Pakistan's president.)

So while the world's media were counting Musharraf out in February, the students successfully forecast that the divide between the PML-N and the PPP would allow Musharraf to hang on for about six months past when they began their study. But even this is but a small part of the big

emerging story played out in advance by modeling key Pakistani policy issues. Figure 10.1 compares the power of only three of many players in Pakistan's political game. Let's see what the picture looks like when we throw
in the main potential threats to Pakistan's civil, secular government. I have
in mind al-Qaeda, the Pakistani and Afghani Taliban, and even Pakistan's
military, with its long history of coups against civilian governments.

Figure 10.2 tells an incredibly distressing story for any who hold out
hope for stable democracy in Pakistan. Pakistan's Taliban and their
Afghan compatriots work together as one, so I present them as if they are
one. Looked at this way, they are far and away the most powerful force
within Pakistan. And al-Qaeda is next in line according to the model, at
least after April 2008, when their power is projected to surpass the government's. Al-Qaeda just continues to grow and grow. Together with the
Taliban they constitute the emerging dominant source of political influence in Pakistan, with only outside influencers like the United States or
the Europeans being possible counterweights. Remember, we are plotting
power—political influence weighted by salience—based on information
known (or at least estimated by my students) back in January 2008 and
not after. Yet here is the headline from the *New York Times* lead story on

FIG. 10.2. Who Will Have the Clout in Pakistan?

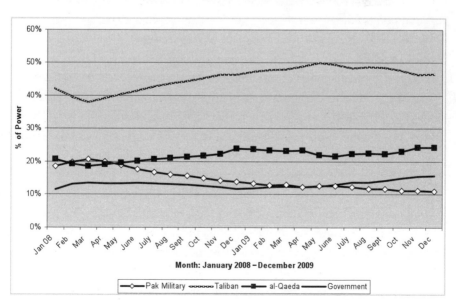

Month: January 2008 – December 2009

Pak Military · Taliban · al-Qaeda · Government

June 30, 2008, months after the analysis was done: "Amid U.S. Policy Disputes, Qaeda Grows in Pakistan." The story goes on to say "it is increasingly clear that the Bush administration will leave office with Al Qaeda having successfully relocated its base from Afghanistan to Pakistan's tribal areas, where it has rebuilt much of its ability to attack from the region and broadcast its messages to militants across the world." My students were able to foresee this troubling prospect half a year ahead of the *New York Times*. Maybe they could have done so even earlier; remember, they only assembled their data in January 2008, when the course, Solving Foreign Crises, began. They saw this result immediately.

There is one other troubling feature to figure 10.2. Other parts of the model's output tell us that al-Qaeda and the Taliban will try to negotiate an arrangement with the PPP and the PML-N. Sharif's PML-N is modestly more open to such talks than is Zardari's PPP. Both prefer to live with the existing status quo vis-à-vis the militant groups while trying to consolidate their own hold on power. In the meantime, the Pakistani military sees itself slowly but steadily losing influence. Such a circumstance raises the prospect that they will try to stem the tide against them by launching a coup to take control of the government. The optimal period for them to take such a step is projected to be between February 2009 and July 2009. Earlier than that they see no need, and later may be too late for them. Pakistan's fragile democracy appears likely to be under assault from the militants who would establish a nondemocratic fundamentalist regime on one side, and from the army that would establish a military government on the other.

What does this mean about Pakistan's contribution to the war on terror? Will they make a more vigorous effort to pursue militants and stamp them out, or will the Pakistani government succumb to the projected growing influence of al-Qaeda and the Taliban? I think you can guess the answer. But just in case you can't, figure 10.3 tells that story.

The status quo commitment to go after al-Qaeda and the Taliban back when my students began their project was at 40 on their issue scale. A value of 40 meant some real efforts to contain the militants but falling well short of trying to stamp them out as the United States wanted. That was equivalent to a score of 100 on the scale. The status quo, with some erosion, was close to the policy predicted to hold, more or less, until the summer of 2008. A position of 0—al-Qaeda's position (not shown)—

FIG. 10.3. Who Will Urge Pakistan's Pursuit of Internal Militant Groups?

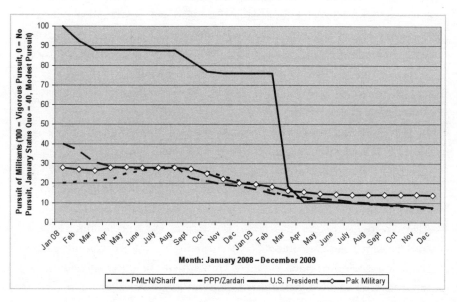

meant "Do nothing against the militants." With that in mind, let's see what we predict for the future.

The dashed line and the dotted line in figure 10.3 show the predicted positions of Zardari's PPP and Sharif's PML-N, respectively. After June 2008, their approach is projected to be little more than rhetorical opposition to the militants with almost no serious commitment to go after them. Talk about a balloon bursting. The air just pours out of the antimilitant effort. That puts responsibility for going after the militants squarely in the United States' corner.

Throughout the remainder of his term, President Bush is adamant (but ineffective) in his commitment to persuade the Pakistani government to go after the hard-liners. After the summer of 2008 even he pretty much gives up on this strategy. Instead, my students found that the U.S. approach will shift ground. The United States' two-pronged strategy of clandestine American pursuit and open Pakistani pursuit of militants will be replaced by a much greater emphasis on the (perhaps clandestine) use of the American military directly within Pakistan. Even that commitment, the student projections indicate, will collapse shortly after the American presidential election. The new president is not likely to do much of anything about the rise of terrorist influence within Pakistan at least through the end of 2009, when the student projection ends. It seems that after a new president is inaugurated—of course, we now know that the president is Barack Obama, but back in the spring of 2008, neither my students nor the model made assumptions about the American election—the war on terror will not be effectively focused within Pakistan.

■ ■ ■

We just had a brief sketch of the major findings regarding Pakistan. There is much more detail that could be discussed about how and why these results arise, but the more interesting question is "What is to be done?" Recall that my students also studied U.S. foreign aid to Pakistan and its likely impact on Pakistani policy. They estimated the current level of U.S. economic assistance to Pakistan in fiscal 2008 as $700 million. That was an approximation, but probably a reasonable one. (The true amount is surely a mix of public and secret information.) They then looked at how congressional and presidential support for that number would change over time, taking into account domestic pressures and the pressures within Pakistan. Here is what they found.

The analysis shows that substantial domestic political pressure is likely to push for cuts in American aid to Pakistan. President Bush and the Democratic Congress are predicted to move apart at least through the summer of 2008 (remember, the data are from January of that year). Indeed, by early summer, the projection was that the president would be pressing for Congress to increase annual aid to Pakistan from $700 million to around $900 million to $1 billion. Did he? Bush proposed shifting an additional $230 million in counterterrorism funds during the summer of 2008. The model predicted that Congress would hold the line on aid during that same period, and they did. Congress has complained in actuality about the amount of aid the United States is giving to Pakistan, but more on that in a moment. After the summer of 2008 the analysis reports that while the president continues to advocate greater aid than Congress supports, the two begin to converge slowly. Both conclude that aid just isn't buying the policy compliance the U.S. government wants.

Put bluntly, American foreign aid is supposed to pay the Pakistani government to go after the militants. It is failing. Indeed, by June 2008 the public discussion coming out of Congress alleged that Pakistan was misusing U.S. aid funds. Money was being spent on items like air defense with Bush's support even though al-Qaeda and its allies in Pakistan are not known to have any air capability. Air defense might be helpful for Pakistan against a threat from, say, India. So, while administration officials countered that the impact of aid on Pakistan's antiterrorist efforts was being underestimated, Congress made the case that the money was being thrown down a rathole.

It is not hard to see that, as projected by my students' analysis, President Obama will face real pressure to support less aid for Pakistan. My students worked through the implications of their assessment and they despaired. Their analysis convinced them that al-Qaeda and the Taliban were getting stronger and that U.S. reluctance to increase aid was more likely to reinforce that trend than reverse it. It was clear from their investigation that the United States cared deeply about getting Pakistan's help in tracking down and neutralizing militants and terrorists operating within Pakistan's borders. It was equally clear that Pakistan's government leaders (the PPP, Sharif and his PML-N, and Musharraf's backers in the military) wanted much more U.S. aid than they were receiving. They could see that the then current U.S. policy did not provide either a sufficient carrot or a painful enough stick to convince Pakistani leaders to put themselves at risk by going after the militants.

With these observations in hand, my students began to think about how they might go beyond predicting developments to trying to shape them (or at least simulate doing so). And so they initiated a search for a strategy that might get Pakistan's leaders to make a more serious effort to rein in the militants. They looked at the possibility of trading aid dollars for policy concessions. Seeing that the Pakistanis want more money and the United States wants greater efforts to track down militants, they wondered whether an aid-for-pursuit deal might not improve the situation from the American and the Pakistani perspectives.

Using foreign aid to secure policy compliance is a time-honored use of such funds even if, at the assumed $700 million in economic aid, it did not seem to be working in Pakistan. Looking at their assessments, my students could see why Pakistan's leaders were not aggressively pursuing militants despite the then U.S. aid program for Pakistan. They could see that the leadership (Musharraf, the PPP, and the PML-N) expected to take too much political heat from al-Qaeda and the Taliban for it to be in their interest at the prevailing foreign aid level. And so my students set out to analyze how that might change if the United States gave significantly more aid than their analysis indicated was going to be the case. Figures 10.4A and 10.4B show the same projections through the end of President Bush's term, but then they diverge. Figure 10.4A continues to forecast the relative influence of the U.S. government and the militants in Pakistan through the end of 2009 if President Obama follows the foreign aid course pursued by the Bush administration. Figure 10.4B assumes that Obama follows the course recommended by my students (and also by his now vice president, Joseph Biden), equal essentially to doubling U.S. aid.

As is evident from the figures, continuing the current aid policy is a losing proposition. With 2008 aid levels, the United States maintains a small power advantage over the militants during Obama's first year in office. That advantage virtually disappears by the start of 2010. The picture is entirely different if the United States and Pakistan strike a deal that trades dollars for aggressive pursuit of al-Qaeda and the Taliban. Rather than coming out of the chute looking to make a deal with the militants, the National Assembly leaders confront the Taliban and al-Qaeda. They impose heavy political and material costs on them and bear heavy costs in return. With the Pakistani government motivated by a doubling of aid dollars, the Obama administration increases U.S. clout in Pakistan at the direct expense of the militant groups (including al-Qaeda, the Taliban, and elements sympathetic to them within Pakistan's intelligence service, the ISI). According to the model's logic and evidence, we can fundamentally change the lay of the land in Pakistan, but to do so, we need to be responsive to the interest Zardari's government has in getting its hands on more money. They won't take the heat against the militants without it. No doubt some of that money will be stolen by corrupt officials, but that's the point. They will want to continue the flow of dollars, and the only way they'll succeed at that is by helping the United States against al-Qaeda and the Taliban.

It is evident from my students' analysis that a promise of greater efforts to go after militants will accompany an increase in U.S. aid. This, however, leaves two questions to be answered: How did we arrive at the idea that doubling aid is the optimal aid-for-pursuit deal, and will each side to such a deal follow through and show that they are really committed to it? My students answered the first question but didn't have enough time to

FIG. 10.4A & B. Reining in Pakistan's Militants by Doubling Foreign Aid

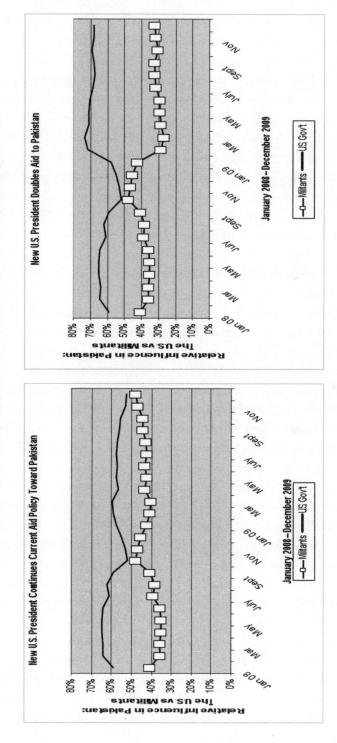

analyze the second, so I will do so here. But first, how do we work out what the optimal aid package looks like?

Back in Game Theory 101, I introduced a way to look at how players resolve trade-offs across issues. In that chapter we were examining the notion of the national interest and saw that there were lots of ways to put a winning electoral coalition together, some of which supported freer trade and others fairer trade, some of which endorsed increased defense spending and others reduced spending. Now, using my students' analysis, we can build on their solutions to the game and apply that methodology— known as win-sets—to their simulated results to evaluate the maximum pursuit that the United States can get from the Pakistani government and the cost in aid dollars to get that maximum pursuit.

Figure 10.5 plots the preferred policies on aid and pursuit of militants held by Pakistan's leadership and the average of the preferred foreign aid package supported by Congress and the president (and their shared view on pursuit of militants) around July 2008 against the predicted status quo on these two dimensions at that time. July 2008—roughly when Musharraf was predicted to be deposed and a new, PPP-dominated regime was expected to take over—is chosen as a prominent early opportunity to strike a deal that trades aid dollars for pursuit of the militants. The solid gray petal-shaped area in the figure shows the range of aid packages and levels of pursuit that are improvements over the status quo from the U.S. government's and the Pakistani government's perspectives. Different points within this gray area show different mutually acceptable trades between aid money and efforts to take down the militants. The optimal deal is in the top right corner of the petal. At that point the United States extracts the maximum effort by Pakistan's leaders against the militants, and the Pakistanis in turn extract the maximum amount of money they can get. More money buys no additional commitment to go after the militants, and greater effort against the militants extracts no additional dollars. That is true because policies outside the petal are not mutually beneficial relative to the status quo, since policies outside the petal are farther from what one or the other set of players wants relative to the fallback position that is the status quo.

So what is the optimal deal? The horizontal arrow in figure 10.5 shows the amount of foreign aid that secures the greatest effort to pursue militants by Pakistan's leaders. That amount is $1.5 billion for 2009. The vertical arrow identifies the maximum effort in pursuing militants that

FIG. 10.5. **Aid Dollars Can Buy Great Pursuit of Militants**

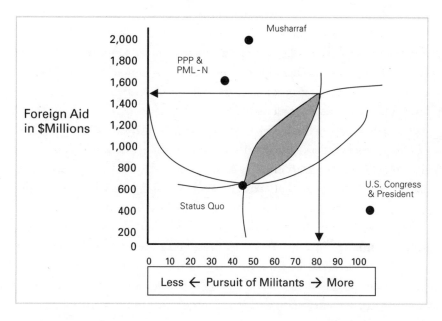

secures $1.5 billion in U.S. aid. That level of pursuit is equivalent to about 80 on the pursuit scale. That is, my students found that for $1.5 billion in aid it is very likely that Pakistan's national assembly, its president, and other important players would really go after the terrorist threat emanating from their country. That is a bit more than double the aid estimated for 2008 and many times the amount advocated by the president or Congress at the outset of the new president's term (according to the model-based analysis). It requires a marked change in U.S. policy.

What would it buy? Put in terms of the issue scale, if the Pakistani leaders get $1.5 billion in U.S. aid, their pursuit of militants will be far above the status quo score of 40 as of January 2008, but it will not equal the intense pursuit identified with a score of 100. That was the level the Bush administration and Congress wanted and surely is the level any American president will desire. So we won't get 100, but we certainly will get a lot more than what is projected without this infusion of money. But please don't get me, my students, or the analysis wrong: the Pakistani government is unlikely to completely quash the terrorist threat just for money. They are no fools. They know that the money will dry up if al-Qaeda and the Taliban are destroyed. So for money they will rein the threat in and reduce it (that's 80 on the scale) but not utterly destroy it (that would be 100 on the scale). For their own political survival they will do whatever it takes. They will try to wipe out the militants if that is their best political path, but if making a deal with the hard-liners looks politically best for Pakistan's leaders, then that is what they will do. A billion and a half dollars would go a long way to convincing them that they are better off going after the militants, insurgents, and terrorists than accommodating them.

Of course, if an aid-for-pursuit deal is struck, each side will have to feel confident that the other side will not renege. Pakistan's leaders must believe that aid dollars will continue to flow, and Congress must believe that, having received the money, Pakistan's leaders will not turn around and still make a deal with the militants. The latter may be of especially great concern because even with an aid deal, al-Qaeda's clout, although diminished, still continues to be substantial, as projected in figure 10.4B. What the model shows is that the leadership in the national assembly will sustain pursuit of militants at around 75–80 on the scale, round after round after round. With the aid deal buying so much greater an effort to combat terrorists within Pakistan, Congress and the president are expected to remain steadfast in their commitment to the proposed aid agreement. All parties to the deal show a real commitment to sustain it, each for their own political benefit.

Without the huge infusion of aid funds from the United States, I'm afraid the projected future is that Pakistan's leaders will make a deal with the militants, who will become a legitimate part of policy making in Pakistan, or that country's government will face another military coup. In all likelihood, American interests will decline and be thwarted. Too bad that this is what's likely to happen. A billion five is a cheap price to pay to stabilize a civil, secular Pakistani government.

IRAN AND IRAQ: IS THERE A MARRIAGE MADE IN HEAVEN?

The future my students projected for Pakistan back in 2008 was grim. The reality a year later is at least as grim. Looking back over their predictions—some about things that have now happened and others about events yet to come—it seems that their play of the predictioneer's game has proven depressingly accurate. Sometimes it really would be better to get things wrong.

Now, in the spring of 2009, we have another chance to test the game. Let's grab that chance and see where it takes us. Just a hop, skip, and a jump away from Pakistan and all its troubles, Iran and Iraq are facing an uncertain future. That uncertainty provides a golden opportunity to dare again to be embarrassed. You see, as I write this in early April 2009, I am again teaching the course on solving foreign crises that was used in 2008 to predict Pakistan's stability (or, more precisely, its instability). Two excellent students in my 2009 course have put my model to good use, looking into the future relations between Iran and Iraq. In doing so, they have uncovered some pretty important insights about likely political developments in each of those countries. Their starting point—*Will Iran and Iraq forge a strategic partnership?*—illuminates a debate at the heart of two entirely different perspectives on American foreign policy: Should the U.S. government keep troops in Iraq, as President Obama proposes to do, or should we pull out altogether? The predictioneer's game can help answer that question.

Let's review a few critical facts before plunging into the analysis. Back before the November 2008 presidential election, candidate Barack Obama promised to withdraw American troops from Iraq within sixteen months. On February 27, 2009, ensconced in the White House, Obama stretched his pre-election withdrawal timetable a little bit, to August 2010. That did not seem to have elicited any great controversy either among "pro-war" Republicans or "anti-war" Democrats. But when he announced the timing of the U.S. withdrawal, he also declared his intention to keep fifty thousand American troops in Iraq. That's no small, token force. It is, in fact, 36 percent of the total number of U.S. soldiers in Iraq at the time he announced his policy. Not too surprisingly, he was subjected to plenty of complaints within the ranks of the Democratic Party for moving too slowly on pulling U.S. forces out of Iraq altogether. Predictably, he also got scant praise in return from the Republicans. Politics is not a warm and cuddly business. Obama took additional heat because these combat-ready troops are slated to stay in Iraq at least until 2011, when a pre-existing agreement with the Iraqi government calls for a full withdrawal. Of course, the possibility remains that the 2011 deadline could be extended indefinitely.

The decision President Obama made in February 2009 and the reality he will face in August 2010 may look alike, and they may not. Pressure within his own party and changing circumstances on the ground might result in a decision to keep far fewer troops in Iraq. But of course it is also possible that President Obama will stick to his guns (fifty thousand of them). I am not going to try to resolve here which he will do, but I am going to use the predictioneer's game to resolve which he *ought* to do. The answer will not depend on my personal inclinations or those of my students; I certainly don't know what they favor. I had barely given this question any thought myself before doing the analysis.

Of course, this investigation touches on only a few aspects of the policy implications of keeping U.S. forces in Iraq or withdrawing them. Facets of American security not examined here may also be influenced by the U.S. decision to pull out of or stay in Iraq. For instance, the troop decision also might make a difference in which way Iran heads in its pursuit of a nuclear capability. But I do not tackle that issue here. I will just say that the prospects of resolving that country's nuclear threat are sufficiently good (based on earlier analyses I have done on Iran) that I do not believe a continued, greatly reduced American military presence will materially tip the resolution of the nuclear issue one way or the other.[1]

WHY MIGHT IRAN AND IRAQ WANT TO BE PARTNERS?

Pulling American troops out of Iraq is predicated on the idea that by the summer of 2010 Iraq will be able to defend itself against internal and external threats to its security. The Iraqi leadership must, of course, be mindful of the giant white elephant on its border as well as the potential of resurgent insurgents at home. One way to cope with its giant neighbor, Iran, is to forge close ties between the two countries. With that possibility in mind, let's think about the range of deals Iran's Shi'ite theocracy might strike with Iraq's secular but Shi'ite-dominated government. As we contemplate a possible Iraqi-Iranian partnership, we must keep in mind that relations between Sunni and Shia Muslims are often extremely fractious, and more so in countries like Iraq, where both groups make up a substantial segment of the population.

Iraq's population is divided roughly 65 percent to 35 percent between Shia and Sunni Muslims, and many followers of the two factions hate one another. That divide certainly was a major factor that gave rise to Iraq's insurgencies and the U.S. creation of those CLCs we talked about in Game Theory 101. During the insurgency, many Shia residents in Sunni areas were driven from their homes, and sometimes murdered on sight, by local Sunni militias. Likewise, Sunni residents in Shia-dominated communities were driven out or murdered. Although things are calmer now and some people have returned to their homes, many have not and animosities linger just beneath the surface, ever ready to explode at the first sign of provocation.

Unlike Iraq, Iran does not have much of a domestic Shia-Sunni problem. That's not so surprising. After all, Sunni Muslims are in scarce supply in Iran. There are about ten Shia for every one Sunni in that country. That is, however, not to suggest that Iranians are warmly disposed or even indifferent to the Sunni branch of Islam. Iran has certainly had more than its share of contentious relations with Sunni-dominated governments in the Middle East and in the wider Islamic world. Most notably, Iran had terrible relations with Iraq during the long years in which the latter was run by Saddam Hussein. Iran and Iraq fought an eight-year war that killed more than a million people and saw the extensive use of chemical warfare. Few in Iraq or Iran have forgotten, and fewer still are likely to forgive, so building bridges between these two countries will not be an easy matter. Staying apart, however, carries its own considerable share of risks.

The Shia-dominated Iraqi government headed by Prime Minister Nuri al-Maliki sees Iran as a potentially sympathetic and like-minded ally. In contrast, he and his closest Iraqi followers may consider their own Sunni brethren a threat to their regime and their vision for Iraq's future. Maliki surely wants to shore up Iraq's security and, according to the data my students assembled, he sees forming a strategic alliance with Iran as the way to do so once the U.S. reduces its military presence or pulls out altogether. His starting position on the partnering issue described in the table on the next page is at 80. That means getting Iran to guarantee Iraq's security. Such an assurance would provide a credible military threat from Iran against any anti-Shia rebellion in Iraq. That could be just the insurance the Maliki government needs.

THE IRAN-IRAQ PARTNER'S GAME

Position	Meaning	Detailed Implications for Iran-Iraq Relations
100	Full Strategic Partnership	Free flow of arms and military technology; a mutual defense alliance; joint intelligence operations
80	Concentrated Partnership	Restricted flow of arms and technology; some intelligence sharing; an alliance in which each guarantees to defend the other
50	Restrictive Partnership	Limited arms flow; no technology transfer; no shared intelligence; each promises not to use force against the other
20	Minimal Partnership	Considerable restrictions on arms flow; no signed alliance agreement at all
0	No Strategic Partnership	No flow of arms or technology; the two governments reaffirm their commitment to the Algiers Accord[2]

Putting such a partnership together, however, will not be easy. Besides the usual complexities behind any international negotiation, it is likely that the U.S. government will present stiff diplomatic opposition to such a move by Iraq. Besides pressure from Obama, we can be confident that those who represent Iraq's Sunni interests will also strenuously oppose any deal with Iran. As for Iran, a deal with Iraq would advance Iran's ambition to become the dominant regional power, but the Iranian government will have to ponder the risks of associating closely with a regime that could fall into Sunni hands. The partnership issue seems especially well suited to evaluating whether the United States is better off keeping some troops in Iraq or removing all of them. Iran, after all, is hardly the state Obama would like to see exercise real influence over Iraqi policy, and a partnership between the two countries could have exactly that consequence.

The table tells us there's quite a range of possible future relations between Iran and Iraq, and of course we need to play the game to work out what is likely to happen. From Barack Obama's vantage point, Iraq ought not to be too quick to jump into bed with Iran. He thinks a policy around 0 on the scale is just right. That is, the Obama administration wants the two countries to go their separate ways while maintaining quiet at the borders, as is their obligation under the terms of their 1975 treaty. But that is not what Prime Minister Nuri al-Maliki wants. He advocates a concentrated strategic partnership (80 on the scale). Maliki's government needs a protector; if it won't be the United States, he would be content to obtain security guarantees from Iran. For him, forging a close association with his much larger neighbor makes a lot of sense. Left to his own devices, Maliki would choose a path that is opposite to the one President Obama wants. Of course, neither Maliki nor any future Iraqi leader will be left to his own devices. There's plenty of pulls and tugs on all sides, so we really do need a tool, like game theory, to help us sort out what the future holds.

While President Obama urges Prime Minister Maliki not to make a deal with Iran, the expert data going into the game indicates that Iran's Ayatollah Ali Khamenei—the supreme leader with a veto over all Iranian policy— welcomes an opportunity for an even closer relationship than Maliki desires. He, too, wants a mutual defense agreement, but he also wants an almost unrestricted flow of intelligence and arms dealing between his government and Iraq's. It appears that Khamenei would like to use Iraq as a base for gathering information about the goings-on in the Arab states next door and beyond. So there is a substantial difference between Obama's vision for Iran and Iraq and the ambitions of the leaders of those two countries.

WHAT WILL IRAQ OFFER TO IRAN?

With this bit of background in mind, we can ask what is likely to happen under two plausible scenarios: (1) Iran and Iraq first work out their respective positions through the normal internal give-and-take of domestic politics and then, having resolved the stance they will take, negotiate against the backdrop of ongoing U.S. pressure in the form of a continuing fifty-thousand-strong American military presence in Iraq; or (2) they each settle their domestic games and then negotiate their future relations bilaterally, without outside interference and with American forces fully withdrawn.

Figure 10.6A displays the evolving positions of four key political figures in Iraq: Prime Minister Maliki; vice president and leading Sunni politician Tariq al-Hashimi; Iraq's President Jalal Talabani, who leads Iraq's Kurdish faction; and Muqtada al-Sadr, the militant anti-American Shi'ite leader. The analysis on which the figure is based assumes that the United States will fully withdraw its troops by August 2010, an outcome favored not only by many Americans but also by many Iraqis. Figure 10.6B displays the same key Iraqi political leaders, but this time having solved the game under the assumption that the United States will maintain fifty thousand combat-ready forces in Iraq. The analyses are not precise about when Iraq's leaders will come to a stable point of view on dealing with Iran, but they do imply that a decision will be reached not much later than August 2010—and quite possibly earlier. The issue appears to be on the back burner for now but it will surely heat up as the U.S. withdrawal date draws nearer.

The model shows that it takes six or seven bargaining rounds before Iraq's political interests (including many more than the four leaders displayed in the figures) come to an agreement on how to deal with Iran. That is a large number of rounds before a stable outcome can emerge. So

FIG. 10.6A. The Likely Iraqi Approach to Iran
If the United States Pulls Out

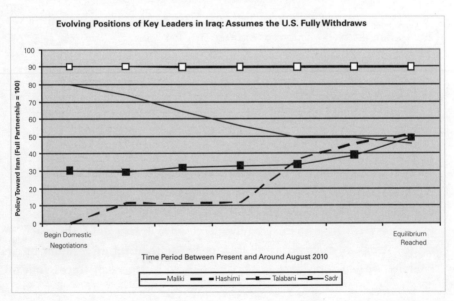

many rounds of internal discussion imply a long stretch of time between when the issue moves to the front burner and when it is settled internally. Apparently it will not be easy for the Iraqis to work out what they want their future relationship with Iran to look like.

Figures 10.6A and 10.6B, viewed together, tell an interesting story. Despite their deep differences, Maliki, Talabani, and Hashimi slowly but surely come around to a collective agreement. They will support a relatively lukewarm relationship with Iran, a relationship not nearly as close as Maliki wants. According to the game, Iraqi diplomats will be authorized to seek an agreement with Iran that includes limited arms flow between the two countries, with no preparedness to transfer technology or share intelligence. In terms of a formal treaty relationship, what is likely to be sought is a promise from each not to use force against the other. That means, in the parlance of international affairs, that Iraq seeks a mutual nonaggression pact. The United States ultimately will support this undertaking, but only after a protracted negotiation. If U.S. troops remain in Iraq, Talabani will feel emboldened to press for an even weaker association with Iran, but he will not prevail. He will go along with Maliki's compromise position if U.S. troops are withdrawn.

FIG. 10.6B. The Likely Iraqi Approach to Iran If the United States Keeps 50,000 Troops in Iraq

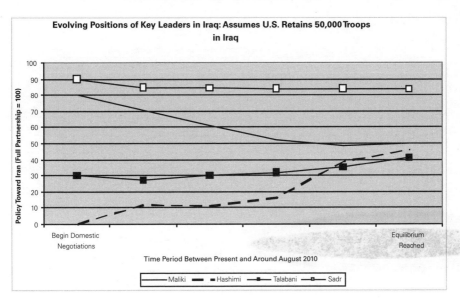

There is one more element in Figures 10.6A and B that is strikingly important. Muqtada al-Sadr, the militant Shia cleric, steadfastly opposes the pursuit of a watered-down, weak partnership with Iran. In the absence of a U.S. military presence, he does not budge from his initial point of view. That perspective favors almost the most extreme partnership anyone advocates. Indeed, as we will see, only some Iranian leaders—like Ahmadinejad—want as much. Sadr advocates a free flow of arms and military technology between the two countries, accompanied by a mutual defense alliance and joint intelligence operations. He will back ever so slightly away from that extreme position if the United States retains troops in Iraq, presumably out of concern for the security of his own operations.

IRAQ'S POLITICAL WINNERS AND LOSERS

Before leaving the internal decision making in Iraq for a look at the comparable domestic evaluation of choices in Iran, we would do well to inquire about who will be Iraq's political winners and losers on this big question of partnership with Iran. Figure 10.7A displays the predicted

FIG. 10.7A. Changing Power in Iraq If the United States Withdraws Completely

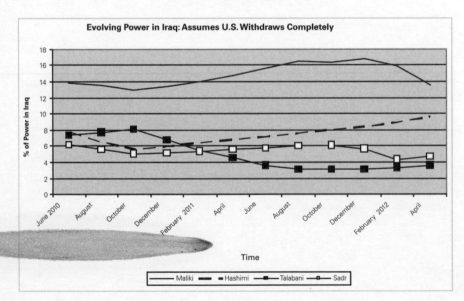

changes in political influence for Maliki, Hashimi, Sadr, and Talabani if the United States fully withdraws. Figure 10.7B evaluates the same power question if Obama leaves fifty thousand combat troops in Iraq.

Even a cursory glance at the projected changes in political power in Iraq suggests that Prime Minister Maliki will need a deal with Iran's Ayatollah Khamenei more urgently if the United States pulls out than if Obama proves true to his word and keeps American soldiers in Iraq. Figure 10.7A indicates that after months of rising political clout, Maliki's influence will start to decline around late spring or early summer 2011. Meanwhile, Hashimi's power will rise steadily. Without a substantial U.S. troop presence, the game indicates that sometime around early to mid-2012 Hashimi will be almost dead even with Maliki in clout. Conversely, if the United States retains a large contingent of combat-ready troops, while Hashimi's growth in power is unabated, Maliki's power does not go into decline. He remains considerably more powerful than his Sunni political rival. Since Maliki has shown himself willing to cooperate with the U.S. government and Hashimi mostly has not, a continued troop presence may be important to prevent Hashimi from becoming a bigger player

FIG. 10.7B. Changing Power in Iraq If the United States Keeps 50,000 Combat Troops There

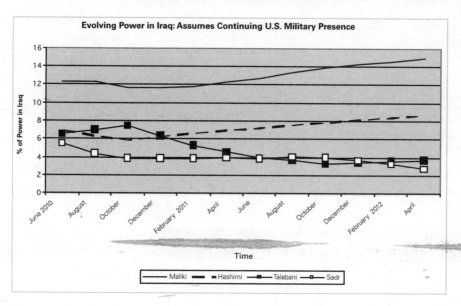

than he already is. Maliki may want to reconsider the 2011 deadline for a complete U.S. troop withdrawal.

With fifty thousand American soldiers in Iraq, Sadr's political future looks much worse than if the United States withdraws. To be sure, the assessment derived from the game indicates that Sadr is entering a period of decline either way, but his downfall is steeper if President Obama resists the pressure to withdraw. President Talabani is also on the way down either way, but he falls faster and farther if the United States pulls out. That is a rather unfortunate combination of circumstances because Sadr is openly hostile to the United States and Talabani views the United States as an important ally.

The really big story in Iraq, however, takes us back to the changing fortunes of Hashimi and Maliki, especially if the United States pulls all of its forces out of Iraq. As I mentioned, Maliki is a reasonably reliable friend. He understands the brazen pursuit of his self-interest; he will just as quickly make a deal with Iran if the United States pulls out as he will make nice to the United States if American troops remain in place. He's got his finger in the wind and he is working out who will do the most to be his guardian angel. The big risk in his political life is being ousted from office; the major domestic threat to his continued hold on power clearly comes from Hashimi. Hashimi wants Iraq to have nothing to do with Iran. Furthermore, he wants to reverse the government's policy of de-Baathification; that is, he wants an end to the ongoing exclusion of former Baathists (Saddam Hussein's party) from the government. And Hashimi staunchly opposes a federal structure for Iraq. Federalism is seen by many—most notably Vice President Joseph Biden—as the most promising means to avert civil war. Thus a political struggle along the Shia-Sunni (Maliki-Hashimi) divide is likely to cast a huge shadow over Iraq if U.S. forces are withdrawn. It is a much smaller shadow with U.S. forces on the scene.

With Maliki's power slipping while Hashimi's rises under the withdrawal scenario, there seem to be only two ways things can go—and neither is good from the U.S. perspective. Maliki can enter into a power-sharing arrangement with Hashimi. That would significantly strengthen the central government and assuage many Sunnis, two good things, but it might also open the door for the Baathists to regain control, a potentially very bad outcome indeed. After all, the projected power in the absence of the United States shows Maliki and Hashimi almost dead even and with

Maliki on a downward spiral while Hashimi is ascending. Maliki, fearful of just such a takeover by Baathists, might opt for the second solution to the threat to his power. Rather than sharing leadership with Hashimi, he might call on Iran to step in and help defend his regime against a nascent Sunni-led insurgency or civil war. That, of course, would be an awful outcome for just about everyone except the Iranian leadership.

IRAN-IRAQ PARTNERSHIP

The feasibility of Iran's army being invited in to help shore up Maliki's regime against a Sunni threat depends, of course, on the nature of the deal the two countries will strike. The internal dynamics in Iran lead quickly—after just three rounds of domestic give-and-take—to a decision on how Khamenei should deal with Iraq in trying to forge a partnership. He will seek a full strategic partnership. Once each country has resolved its own views on partnership, it will be time for the respective negotiators to come together to discover whether they can find common ground for a deal.

Figures 10.8A and B show what is likely to emerge from bilateral Iran-

FIG. 10.8A. Iran-Iraq Negotiations After the United States Withdraws

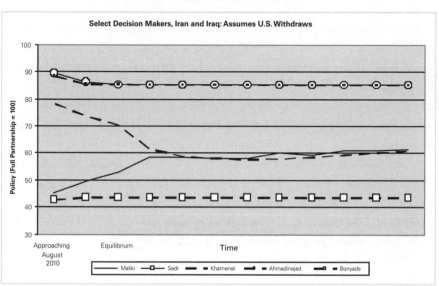

FIG. 10.8B. Iran-Iraq Negotiations If the United States Does Not Withdraw

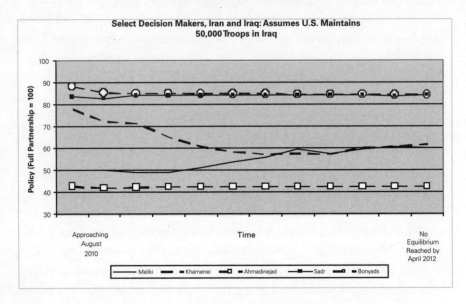

Iraq negotiations if the United States has pulled out militarily or keeps fifty thousand troops on the ground. The pictures tell radically different stories. Without U.S. troops present, Maliki and Khamenei quickly come to terms. If American forces are on the scene, it looks like the negotiations will be abandoned—or at least tabled—well before an agreement is reached. Indeed, the game suggests that the two governments will not have come to terms with each other even after more than two years of negotiations, *if and only if* Obama maintains a fifty-thousand-strong combat contingent in Iraq.

Figures 10.8A and B depict the policy positions of the two principal decision makers—Khamenei and Maliki—during the course of bilateral negotiations, but the figures also show the evolving policy stances of the most extreme elements with real clout in each country. Thus we see the near polar-opposite positions of Iran's President Ahmadinejad and the Bonyads, a group of Iranian tax-exempt charities that exert massive control over much of Iran's economy and have enormous influence over Khamenei and the ruling council of Ayatollahs. Khamenei appears comfortable with showing real flexibility to advance the prospects of striking a deal with Maliki's government whether U.S. troops stay or go. But Mahmoud Ahmadinejad resists the partnership agreements that could be in the cards regardless of whether the United States withdraws.

According to the predictioneer's game, Ahmadinejad starts out and ends up advocating much firmer Iranian influence over Iraq than the Iraqis can agree to, and in doing so he is likely to alienate Khamenei.[3] Indeed, as we will see, Ahmadinejad doesn't get his way and this gradually costs him political influence. Meanwhile, the Bonyads—that is, the principal moneyed interests in Iran—remain equally steadfast in their opposition to what could be a very costly Iranian partnership with Iraq. They hold out for about as weak a set of ties as the United States is willing to live with. They advocate a bit more than cordial relations between the two countries, but not much more than that. Who knows, as the Ayatollahs' influence declines—and as we will see, it is already doing so—the Bonyads may become a vehicle through which the United States can find common ground with important stakeholders in Iran.

On the Iraqi side, Muqtada al-Sadr plays much the same part that Ahmadinejad plays in Iran. Sadr too proves all but immovable. However, even as he and Ahmadinejad try to scuttle an agreement, the game indicates that if the United States withdraws, Maliki and Khamenei will swiftly arrive at an agreement. The deal they are predicted to strike if Obama withdraws all American combat-ready troops is at 60 on the issue scale. This means the two countries will engage in a fair amount of arms transfers. They will capitalize on some coordination between their intelligence services and they probably will sign an alliance (such as a mutual entente) that assures more than nonaggression between them but that does not go so far as to provide guarantees of mutual defense. Such an arrangement probably would be sufficient for Maliki to call on Iran to defend his government against a Sunni uprising if one were to occur, thereby improving the odds of keeping Hashimi at bay.

If, however, the United States keeps fifty thousand troops in Iraq, the picture is entirely different. As can be seen in figure 10.8B, although negotiations can result in Khamenei and Maliki coming to terms, the conditions for a stable outcome are not present. That is undoubtedly because Maliki will face great political pressure at home. That pressure will oppose his signing a partnership agreement with Iran. So, facing such stiff domestic political pressure, Maliki will put any possible deal on hold. Even after simulating more than two years of negotiations, the model does not arrive at an equilibrium outcome: the game goes on. According to the game, the discussions would most probably be broken off well before the two sides could discover a deal the Iraqis could sell politically at home. That is, the American military presence is sufficient to hold Maliki's feet to the fire, keeping him from making big concessions to Khamenei. There are ways to overcome the problems Maliki will face, but considering that there is a reasonable chance that Iraqi or Iranian diplomats might read this, I leave it to them to work out how to solve their problem. It isn't likely that they would listen to what I have to say, but why test those waters?

Before closing this opportunity to be embarrassed, let's take a look at the predicted evolution of political influence in Iran. This reveals some interesting insights that may make us more hopeful for the future, especially if the United States keeps forces in Iraq long enough to buy time for the predicted developments in Iran to take hold.

Figure 10.9 shows the projected changes in political power among four key Iranian interests. Ayatollah Ali Khamenei, the most powerful figure in Iran, is, according to projections from the predictioneer's game, entering a long period of political decline, probably to culminate in his retirement. This signals a major change in Iranian affairs because he has a veto over virtually all policy decisions. Less well known in the west are Major Gen-

FIG. 10.9. Evolving Power in Iran: Some Hope for the Future

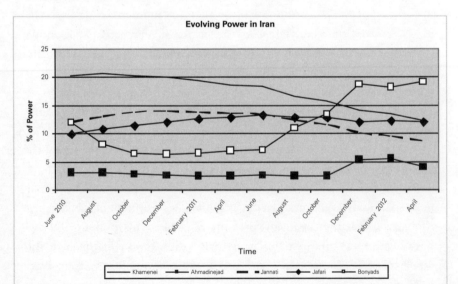

eral Mohammad-Ali Jafari and Ayatollah Ahmad Jannati. Ayatollah Jannati is chairman of the Guardian Council. An antireform cleric, he can veto candidates for parliament, and he has the authority to assess whether parliamentary decisions are consistent with the constitution and with shariah (that is, Islamic) law. As such, he is nearly as powerful a figure as Ayatollah Khamenei. Major General Jafari commands the Islamic Revolutionary Guards Corps, the elite military unit whose support sustains the regime. Each of these individuals is far more powerful than Iran's president and American nemesis, Mahmoud Ahmadinejad. Figure 10.9 also shows the evolving power of Iran's Bonyads. The Bonyads were originally created during the era of the Shah and then were completely recast after the Islamic Revolution in 1979. As I have noted, they control vast sums of money, are exempt from taxes, and answer to no one except Iran's Supreme Leader, Ayatollah Khamenei. They control Iran's purse strings and, not surprisingly, have a reputation for corruption and mismanagement. But they are also political pragmatists, as we have seen, so they may be people with whom the Obama administration can find a path through to resolve many of the tensions between the United States and Iran.

Figure 10.9, if correct, tells a startling story of emerging change in Iran.[4] Key religious leaders like Khamenei and Jannati are in political decline. Other clerics (especially the less politically involved Qum clerics, sometimes known as the Quietists) may be picking up some of the lost power, but the lion's share of shifting political influence falls into the hands of Major General Jafari, as the principal representative of Iran's elite military units, and the Bonyads, Iran's money managers. Business interests are also gaining in influence. That is, a more secular and pragmatic, albeit militarized and corruption-laden regime appears to be emerging as Iran's theocracy goes into political, if not spiritual, decline. While the theocracy is likely to hold on to the symbolic trappings of power, real control is slipping away from them and toward a more conventional strongman, moneyed dictatorship.

The analysis certainly suggests different worlds depending on whether American troops are in Iraq beyond August 2010. If Iran can strike a significant partnership arrangement with Iraq, as is likely if the United States withdraws, Iran will be well on its way to asserting itself as the dominant power in the region and will position Iran to resume its aggressive efforts to export its form of fundamentalist Islam. Such a triumph might even reverse the Ayatollahs' slide toward lost political control. Fortunately, these developments are unlikely because President Obama is likely to be a man who sticks to his word. Keeping U.S. forces in Iraq appears to be sufficient to deter the strong ties between Iran and Iraq that would provide a basis for Iranian military intervention to defend the Maliki government against a potential Sunni uprising. And it also appears to be sufficient to keep Maliki strong enough that Hashimi is likely to think twice before attempting to push him aside. With a continued U.S. military presence, time can be bought to provide the opportunity for a less anti-U.S. regime to take control in Iran, a trend that emerges in the game under such conditions. Staying creates the chance to deal with a "normal" petty dictatorship and maybe, just maybe, even a nascent more democratic regime. Withdrawal raises the prospects of helping the Ayatollahs stay in power while also jeopardizing the prospects of a pro-U.S. Iraq in the future.

With our analysis in hand, we can cast the debate over whether to leave the troops in place or pull everyone out by August 2010 in a clearer light. Pulling out is tantamount to inviting Iran to step in to fill the void. That would be extremely dangerous from the American perspective. On the flip side, however, we should also ask to what extent a continued U.S. presence is likely to stymie efforts by President Obama and his secretary of state, Hillary Rodham Clinton, to negotiate a resolution of Iran's growing nuclear threat. The risk, based on other assessments I have done, seems small. What is more, if the Qum Quietist Ayatollahs are on the rise, as seems true from the game, and if the Bonyads and the military are also on the rise, then Iran is soon to enter a more pragmatic era that will help foster resolution of issues, like the nuclear one, that loom so large now. Time will tell. I invite others studying these problems from different perspectives to dare also to be embarrassed and tell us now what they think will happen two years into the future.

11

■

THE BIG SWEEP:

THE HISTORY OF WORMS,

OR BALI HIGH, BALI LOW

PERHAPS A GOOD place for us to close is with an example that looks way back in time and one that projects way forward in time. First, I'll evaluate a problem that is almost nine hundred years old, using only information possessed by the relevant decision makers at the time—popes and kings—to illustrate how game theory can capture and predict tectonic shifts that alter society over the course of hundreds of years, in this case the period between 1122 and 1648. Then we'll take a look at ourselves one hundred to two hundred years from now.

But first, let's go back to 1122 and see how we could have predicted then the essential end of the Catholic Church 526 years later.

■ ■ ■

The Catholic Church is not what it used to be. In the good old days, especially before 1648, the Catholic Church was to European politics what the United States is today: the hegemonic power, the big cheese, the *capo di tutti capi*. That was not always true. Before the tenth or eleventh century, it was big, but not that big. The pope was the bishop of Rome and not much more. Between about 1087 and 1648, the political clout of the Church rose and then fell. This happened, in my opinion, mostly because of a deal it struck with the Holy Roman Emperor in 1122 and with the kings of France and England at about the same time.

The pope has a great job. He lives in terrific digs right in the heart of

Rome. He has a fabulous art collection, the best supply of Italian food anyone could want, and even his clothes, we must admit, are pretty cool—his
hats are out of this world. He travels wherever he wants in grand style, and
he is adored by millions of people. Still, it's not as good a job as it once was.
The best time to be pope was between the papacies of Innocent III (1198–
1216) and Boniface VIII (1294–1303). Those were the days when popes
really had it all—fame, glory, riches, sanctity, power. After that, they went
into a long downward spiral, punctuated most noticeably by the Treaty of
Westphalia that ended the Thirty Years' War.

The Treaty of Westphalia in 1648 formally made kings sovereign within
their territorial boundaries. They could choose (or let the people choose, an
idea whose time had not yet come) the religion for their domain. The closer
a kingdom was to Rome, the more likely it was to remain Catholic. As one
got farther from the pope's reach, however, Protestants did better. Westphalia enshrined the idea that foreign powers should not interfere with any
country's internal policies. This really limited the Catholic Church's ability
to dictate policies as it had done for centuries. Although the Treaty of Westphalia made these points explicit, they had been in the works for a long
time. And the real action that made such conditions feasible started at least
five centuries earlier, when the sale of bishops' positions was resolved by the
Concordat of Worms.

Historians have a quite different take on the development of modern
sovereign states than I do. My view is shaped by the game that was set up
at Worms. The standard account is that the Catholic Church promoted
economic growth (banning usury only because it was sinful) and managed
reverent kings who deferred to the popes' choices of bishops, which gave
the Vatican localized control over much of Europe. My perspective is that
the Church actively tried to hinder economic growth in the secular realm
and that kings only really deferred to papal choices for bishops where and
when they were forced to for economic reasons. My view looks at the
Church as a political power more than as a religious institution. Please understand, I am not questioning the sincerity of Catholic religious beliefs
today, or at any time in the past. I am just recognizing that in addition to
its religious mission—maybe because of it—the Catholic Church played
power politics.

I contend that, eventually, economic growth made the pope all but irrelevant politically, and it was that very growth—and the contest for it—

that made the terms of the Treaty of Westphalia ultimately possible. I will show how each of these developments was dictated by the strategic implications of Worms and, therefore, that each was predictable. Worms established a way for popes to sustain substantial power for a long time, but it also made inevitable that the Church would ultimately become subservient to the state. In that sense, the pope of 1122 (Calixtus II) did what was good for him and his immediate successors, but at the price of selling out the political prospects of popes centuries later.

The agreement reached at Worms resolved the investiture struggle over bishops. Before Worms, the Holy Roman Emperor and Catholic kings sold bishoprics within their domain. Naturally, the pope objected to this practice. He wanted greater control over bishops. They were, after all, supposed to be his emissaries. Under the concordat, the pope gained the right to nominate bishops, and the king the right to approve or reject the nominees. When a new bishop was installed, the king gave up control over the symbols and trappings of the bishop's office, including its income. In exchange, the bishop promised military assistance and loyalty to the king as sovereign of the territory occupied by the bishopric. In this way, the king transferred back to the church, and to the bishop as its agent, the right to the tax revenues from the see. During the vacancy between the death of the old bishop and the consecration of the new one, the revenue from the bishopric went to the king. This revenue could be substantial. The longer the bishop's office remained vacant, the longer the king received the revenue instead of the church. But rejection of a papal nominee was bound to irritate the pope, and that could be politically and socially costly to the king. The pope could excommunicate the king, or he could interdict the bishopric. That meant that no one in the bishopric could receive any of the sacraments. This was tantamount to fomenting civil war against the king in that deeply religious age.

The king's right to the see's income during a vacancy represented a property right that belonged to the king as sovereign over the territory of the see, and not to the king as an individual. The king could not sell the future right to control the regalia, nor could this right be inherited except by ascent to the throne. The right belonged to the king's successor, who might be his child or might be from an entirely new line. The king held the right to this income, then, as the kingdom's agent and not as his personal, private property. This was a significant departure from feudal practice. It established the sovereign claims of the monarch on behalf of his citizen-subjects within the territory of each bishopric in his domain. It was the beginning of the state as we know it.

Although the actual game set up at Worms is a bit more complicated than the model I present, the game tree in figure 11.1 is close enough to capture the essentials. The pope chooses to nominate a bishop. The nominee is either someone especially to the pope's liking, or someone more to the liking of the king. The king, in turn, can agree to the pope's nominee or reject him. If the king rejects the nominee, then he earns more money but annoys the pope, who must then nominate someone else to be bishop. If the king agrees to the pope's nominee, then the king earns less money because the bishopric does not remain vacant for long, but he improves his relations with the pope. So as long as the pope and the king agree on a nominee, both benefit, although in different ways. On acceptance of a bishop, earnings from the see benefit the pope, and, depending on who the bishop is, the pope either has a loyal ambassador, or the pope relies on a person more loyal to the king than to the pope, a potential fifth-columnist within the pope's circle. On agreement, the king has an ac-

FIG. 11.1. The Game Set Up at Worms in 1122

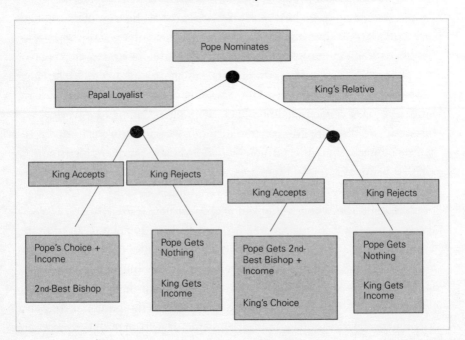

ceptable bishop to work with, and if there is some delay between the death of the previous bishop and the installation of the new one, the king also gets some extra income. Figure 11.2, on the next page, shows a simple version of this game.

The pope's choice looks pretty easy to make. If he nominates someone expected to be loyal to him—a relative or a member of his papal court—and the king accepts, the pope gets his best choice as bishop and he gets the income from the see, eliminating the vacancy as quickly as possible. The problem is that the king might say no to this proposal. Okay, you think, so maybe the pope should nominate someone loyal to the king. Then at least he gets the income. But the king could turn that offer down too. Here lies the kernel of the undoing of the Catholic Church 526 years later. Let's look at the king's choices with a numerical example to illustrate how this works. The scale of the numbers is not important here as long as the order of the size of the value to the pope and to the king under different conditions is right.

Setting aside for the moment the matter of income, let's say that the pope values a bishop who will be loyal to him at 5 points and a bishop loyal to the king at only 3. Not getting the king to agree on a bishop at all is worth 0 points to the pope. This order of values makes clear that the pope prefers his own guy to the king's man, but he prefers the king's choice of bishop to no bishop at all. Let's say that the king places a value of 5 on a bishop who is related to him and a value of 3 on a bishop whose loyalty is expected to be with the pope. The value of having no bishop at all is 0 for the king, just as it was for the pope.

Now comes the fun part. How much is the income from a bishopric worth? I will assume that a poor diocese produces an income worth 1 additional point for whoever gets the income. A moderately wealthy see's income is worth 4 points, and a really rich bishopric produces an income of 6. The game tree shows the pope's benefits first and the king's below the pope's at each place where the game tree ends.

In Chapter 3, when we looked at the banker's game (Paris or Heidelberg), I promised a more interesting game later, and here it is. To solve the game, the pope has to look ahead to figure out what the king is likely to do. The king's decision depends very much on how valuable the income is from the bishopric. When the diocese has a small income—just 1 extra point—the king can get 3 points by agreeing to the pope's choice of a bishop loyal to the papacy, but only 1 point—the value of the income—by

FIG. 11.2. A Numerical Example of the Game Set Up at Worms

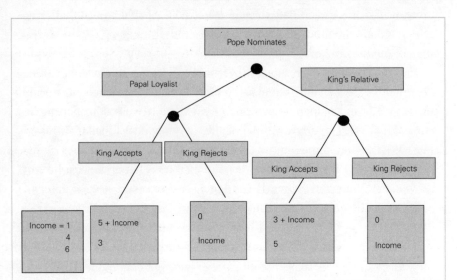

rejecting that nominee. Now, the king would of course do even better if the pope chose one of the king's relatives to be bishop—then the king would receive 5 points in value by accepting the nominee. The Concordat of Worms, however, stipulates that the pope moves first, nominating a candidate to be bishop. The pope has worked out that if he nominates someone loyal to him he will get 6 points, 5 for the bishop and 1 (in the case of this poor see) for the income. That is more than he can get by nominating a relative of the king. So in a poor bishopric, the pope picks someone he likes, forcing the king's hand. The king agrees to the pope's choice, and all is right with the world. This is the world that historians think prevailed because they noticed that kings almost never turned down a nominee to be bishop.

The historians, however, are mistaken, in my opinion. If we look at differences in the income from dioceses in France, for instance, during the reign of Philip Augustus (1179–1223), we discover that the pope overwhelmingly chose people from his own court in poor dioceses, but he chose relatives of the king in moderately well-to-do bishoprics. That is exactly what the game set up at Worms back in 1122 leads us to expect, and that is the key fact that will make the pope want to limit the economic growth that kings can tap into.

When the income is worth 4 points instead of 1, the king gets more value by rejecting a candidate to be bishop who is a papal loyalist than from agreeing to the nominee (4 points versus 3). But if the pope chooses a relative of the king to be bishop, the king is better off saying yes to the candidate than saying no, even though this means giving up the bigger income. The king receives a value equal to 5 points if his relative becomes bishop and only 4 from the income earned in a moderately well-off bishopric. So in wealthier sees, an attentive, politically savvy pope switches his strategy and gives the king someone the king wants. Then the pope earns 3 points for the choice of bishop and another 4 in income. That is the best he can do when he knows the king has an incentive to reject a candidate for bishop that the pope really wants.

Now imagine a really wealthy diocese where the income is worth 6 points. The income from the diocese is worth even more to the king than getting along with the pope. The Church can no longer compete with the monarch for political control—the king just doesn't care about bishops anymore. He just wants the income, and so he rejects any and all nominees. The Concordat at Worms breaks down and we are in a new world in which kings keep incomes in their territory and popes can pick whomever they want as bishops. That, of course, is essentially the situation for the modern Catholic Church. It remains a major religious body, but it is not a major political-military player.

We can see that Worms set up a system that creates some interesting incentives. The higher the income from a see, the harder it is for the pope to get his preferred candidate as bishop. Valuable sees require the pope to make sacrifices. He has to agree to a bishop who in a pinch is more likely to support the king than the pope—or else the pope loses income. This gives the pope a reason to stymie economic growth outside the Church's domain. In fact, shortly after 1122, the Church adopted a series of new programs that were likely to hinder economic progress outside the Church. Was that a coincidence? We cannot know, we can only see that changes introduced by the Church—and by kings—after 1122 were consistent with their new incentives created at Worms.

For example, in the First Lateran Council (1123), a gathering of Church leaders to agree on important new rules, we can see that the role of celibacy for the clergy was made more important. The council prohibited the clergy from marrying or having concubines. The Church was clear about the motivation behind the stricter celibacy rules. They seem not to have been designed to promote purity so much as to clarify the Church's claims on property. The new rules improved the odds that the property of the clergy would belong to the Church rather than to heirs. At the Second Lateran Council (1139), more serious changes were instituted. This council dealt with questions of inheritance and usury. On the inheritance of the private property of deceased bishops, the Church raised the stakes and ensured that they, and not any secular venue, would be the beneficiary:

> The goods of deceased bishops are not to be seized by anyone at all, but are to remain freely at the disposal of the treasurer and the clergy for the needs of the church and the succeeding incumbent. . . . Furthermore, if anyone dares to attempt this behaviour henceforth, he is to be excommunicated. And those who despoil the goods of dying priests or clerics are to be subject to the same sentence.[1]

By imposing excommunication on violators, the Church raised the risks that families or monarchs ran in trying to seize the "personal" property of deceased clerics. That put the money right where the Church wanted it: in its coffers.

The council went on to make usury "despicable and blameworthy by divine and human laws," and cut off usurers from the Church, depriving them even of a Christian burial unless they repented. "Usury" in those days just meant lending money with the expectation of making a profit, not necessarily a big profit as the term implies today. Before 1139, usury had been forbidden to the clergy, but it had not been elevated to a mortal sin for ordinary people. The effect of banning moneylending for profit was to raise the price of money and to create a potential shortage of would-be lenders.

Just as modern-day central banks increase interests rates to slow growth, so the twelfth-century Church raised interest rates by denying heaven to those who lent money for profit. Though the Church used scripture to justify its action, there was a widely held view among Church canonists—the Church's lawyers—that there was no doctrinal ban on usury in early Catholic teachings or scripture. As they noted, Jesus threw the moneylenders out of the Temple. He thought it wrong to engage in such business in the Temple, but he did not argue that the business was wrong. He just wanted it taken outside. And of course we should remember that the Church had not used scripture to ban usury in its preceding thousand years.

Enforcing the ban on moneylending for profit (and why else would someone lend money?), however, proved to be a difficult problem. The sin of usury required intent: the moneylender had to intend to make a profit, so whether the moneylender did or did not actually make a profit was beside the point. The Church recognized how hard it was to determine whether a lender *intended* to make a profit.

To deal with intent, the Church wisely shifted enforcement from its lawyers to its theologians. They reasoned that while human law might fail to recognize a usurious loan, God knew whether a lender intended to make a profit no matter what subterfuge was used to mask the return. Therefore, anyone having made such a profit and failing to make restitution or to show sufficient contrition before death was condemned eternally.

To facilitate restitution for usury, and to heighten the threat of damnation associated with lending, the Church established new institutions. The Fourth Lateran Council (1215), for instance, made annual oral confession mandatory. The Church distributed confessor's manuals with specific instructions for dealing with merchants and others likely to have engaged in usury. Through the confessional, the Church provided a means by which those otherwise damned for usury could save their souls. The path to forgiveness, however, generally included making financial restitution to the Church rather than to those from whom a profit had been made. While the threat of eternal damnation was powerful indeed in the twelfth century, still, moneylending continued.

Of course, merchants and others continued to devise clever schemes to hide what they were doing. They not only manipulated exchange rates but also wrote misleading contracts and created false stock companies (sound familiar?) to hide their true financial arrangements. The risks (including eternal damnation) had been raised, so naturally the expected rate of return had to rise too. The consequence was to make loans costlier and thereby to diminish money for investments and growth relative to what it otherwise would have been.

If we look at last wills and testaments after 1215 we discover a great increase in deathbed confessions of usury. These confessions were frequently accompanied by wills that make penance by leaving money to the

Church to make up for the dying sinner's usurious past. This was a great boon to Church income and took a great deal of money out of the secular economy. It must have left relatives pretty unhappy, but how could they argue against the greater good of eternal salvation?

■ ■ ■

During the twelfth century the Church also adopted new rhetoric that limited growth even as it promoted wealth within the Church's lands. At the same time that new mendicant orders like the Cistercians built and operated wind and water mills and other labor-saving devices to improve efficiency, the Church began to promote the view that idle hands are the work of the devil. The Church discouraged the spread of machines and other labor-saving technology in the secular realm (though not within its own domain). Certainly by discouraging the spread of labor-saving technology, the Church was reducing labor productivity and thereby economic growth in the lay sector. This, as we saw, could only improve the pope's chances of getting the bishops he wanted.

Kings, of course, did not sit idly by while the pope worked to strengthen his hand and weaken theirs. The decades immediately after the Concordat of Worms saw a dramatic flowering of political institutions in England, in France, and elsewhere on the Continent. Whether intentionally or not, many of these new institutions and programs challenged the pope's influence and secured a higher economic growth rate for the king's subjects, and thus higher tax revenues for the king—higher, that is, than they would have been if the pope's policies had been left unchallenged. Consider, for instance, the series of legal reforms introduced by Henry II (you know, *The Lion in Winter*) in England during the mid-twelfth century. These reforms became the foundation of English common law.

Henry moved to protect property rights and rights of inheritance. These actions made it easier for peasant families to predict whether they would continue to benefit from the land they worked after the head of the family died or whether the lord of the manor would take away their opportunity to farm the land. Henry's writs greatly shortened the judicial process for determining rights of access to the land and made for a more smoothly operating agricultural system. His new rules proved highly popular and effective in securing the property rights that are essential to economic growth, and they enhanced the king's credibility as the person who would ensure order and justice in matters of property. Before the writs, tenant farmers were reluctant to invest in their land, but once Henry's writs protecting property rights were in place, more effort was made by those working the land to produce more—for their own benefit as well as for the benefit of the lord of the manor.

Henry did not limit himself to improving property rights for ordinary people. He also acted to impose restrictions on Church rights, a bold move indeed. Through a writ called *utrum,* he asserted the king's primacy in determining whether a dispute belonged in his secular courts or in the church's ecclesiastical courts. Prior to Henry there was a presumption in favor of the jurisdiction of the ecclesiastical courts. *Utrum* reversed this presumption among litigants in England, a presumption that had been in place since the time of William the Conqueror a century earlier. Finally, Henry also moved to protect the patronage system that gave landowners the benefit of choosing clerics for appointment in their personal churches. This ran directly counter to the efforts of the church to take all such influence away from the secular domain, as is evident from the rulings in Lateran II (1139) and III (1179).

Henry's effort to strengthen his hand against the church was further reinforced by his use of the jury system to replace trial by ordeal. Trial by ordeal decided innocence or guilt through the presumed intervention of God. Two common ordeals, both supervised by the Church, involved submersion of the accused in deep water or forcing the accused to hold a red-hot piece of iron for a prescribed amount of time. Failure to stay submerged for the prescribed time was taken as proof of guilt, as was the inability to hold the red-hot iron. As Henry's legal adviser observed at the time, guilt or innocence had more to do with the thickness of one's calluses or the ability to hold one's breath than anything else.

The shift to a jury system helped weaken Church institutions and income. Trials by ordeal were supervised by the clergy, who were well compensated for their participation. For instance, two priests are known to have been paid ten shillings for blessing the ordeal pits near Bury St. Edmunds in 1166. At the time, a worker's daily wage was about one penny, and a villein and his entire family could be purchased for 22s.[2] So 10s for a blessing was a significant amount of money. At 12 pennies to the shilling, the blessing cost the equivalent of 120 days of labor for an ordinary worker. That amounts to over $5,600 at the current U.S. minimum wage just to get the ordeal pit blessed (it's about double that when evaluated relative to the income of the average American worker rather than those earning just minimum wage). With the stroke of a pen, Henry cut Church income substantially and increased his own by introducing the jury system.

The administrative structure of the modern state also began to emerge in the twelfth and thirteenth centuries. The king's right to levy taxes for reasons other than necessity gradually developed in exchange for political concessions to his subjects, most notably in 1297 when Edward I accepted Confirmatio Cartarum. Edward, in "confirming the charter," acquiesced to the changes that had been wrought by Magna Carta eighty years earlier (King John had agreed to and then promptly reneged on the deal). These new tax revenues gave the king the ability to muster an army without having to rely on the intricate rights and restrictions implied by the feudal order. Consequently, the bishop's military guarantee to the king "through the lance," granted at Worms, diminished in importance. The pope, in contrast, continued to rely on feudal commitments to raise an army.

King's courts at fixed locations in England and in France replaced the itinerant justice of an earlier time, thereby centralizing judicial control in the hands of the king, further diminishing the role of the Church as an adjudicator of disputes and further emphasizing the king's territorial sovereignty. Additionally, kings began to claim that they ruled by divine right, thereby challenging the pope's special position as allegedly chosen by God. In the jockeying for control, both kings and church evolved new institutions and methods to foster or stymie economic growth and wrest political control.

The eventual result of all of this competition was just what the game set up at Worms predicts. The Church worked to keep income high within the ecclesiastic domain and low elsewhere. Kings worked to achieve the opposite, seeking control over courts and taxes wherever and whenever possible. Eventually, secular wealth became so great that, as the game implies, kings stopped caring who the bishops were. Kings no longer felt a need to keep the pope happy, and the dominance of the Catholic Church was replaced by the dominance of sovereign states in a secular world.

■   ■   ■

Now that we have seen how we could peer ahead to the big picture five hundred and more years after 1122, let's do the same for our own time. Let's take a look at the inconvenient truth that won Al Gore the Nobel Peace Prize.

The world seems to be undergoing significant warming. The rise in temperature is melting ice in the North Atlantic and elsewhere, raising ocean levels and threatening to sink low-lying island nations and mainland coastal areas. The gathering of more ferocious storms promises years of destructive forces from wind and rain. Higher temperatures will push some temperate climates into the subtropical zone and some subtropical environments into the great stifling heat of the tropics.

First, let me provide a little background on the issue of global warming. After years of debate, including warnings in the 1960s and 1970s of an approaching new ice age, there now seems to be broad agreement within the scientific community that the earth's temperature is on the rise. How much of the rise is due to human activity and how much to a normal cycle in earth's climate is less easily agreed on, mostly because the cycle seems much longer than available weather data. We know, for example, that the High Middle Ages (what we used to call the Dark Ages) were a warm period accompanied by rapid economic growth. We know things got colder, at least in Europe, roughly from the Renaissance until probably sometime in the nineteenth century. And we know things are getting warmer again. We also know the increase in temperature is larger than seems to have been true at any time over the past thousand or so years.[3] Of course, in the earth's history a thousand years is a short time, though on the human clock it is pretty long. Scientists seem to concur that the extra rise in temperature is associated with industrialization and modern chemical-fertilizer farming, and that fossil fuel use is among the big culprits. Obviously, concern about global warming is strong enough that there are ongoing international efforts to bring it under control and even reverse it.

What can game theory tell us about efforts like Kyoto, Bali, and Copenhagen—that is, international conferences to regulate greenhouse gas emissions—to find solutions to the real threat of global warming? What can we learn that will help us carve out a better future for our species and our environment in the centuries to come? Frankly, we will see that agreements like the Kyoto Protocol and the efforts at Bali or Copenhagen to reduce greenhouse emissions, especially carbon dioxide emissions, are not likely to matter. They may even be impediments to real solutions. That is not to say that there is not good hope for the future. There is, because global warming produces its own solutions.

Back in December 1997, 175 countries—not including the United States—signed the Kyoto Protocol. The Kyoto signatories agreed to a benchmark year, 1990, against which to establish targets for greenhouse gas reductions. Greenhouse gas emissions had been rising dramatically in the years between the benchmark and the agreement. The protocols call for a 5.2 percent reduction in greenhouse gas emissions relative to their levels in 1990. That is about equivalent to saying there should be nearly a 30 percent reduction when compared to the then expected emissions levels in 2010. Some signatories were called upon to make much greater sacrifices than others. For instance, the European Union agreed to reduce its emissions by 8 percent. The United States was asked to reduce its greenhouse gases by 7 percent, Japan by 6 percent, and so forth. A few countries, such as Australia, were given permission to increase their emissions. No restrictions were imposed on developing countries like India and China (or Russia, assigned a 0 percent reduction), although they are now among the world's largest greenhouse gas polluters. The United States declined to sign on because it objected to the exemption given to rapidly growing economies like China's and India's.

Kyoto produced a large market in which polluters and nonpolluters could buy and sell "pollution rights." This market has helped to rationalize decisions at the level of individual firms, but it alone has so far failed to result in the magnitude of reductions envisioned by the Kyoto Protocol. As we will see shortly, enforcing the 1997 agreement has been virtually impossible.

One consequence of the difficulties encountered since 1997 was a meeting in Bali, Indonesia, in December 2007. The Bali meeting had more modest goals than Kyoto. It represents an interim step on the way to a 2009 deadline by which it is hoped there will be a new international agreement in Copenhagen. After considerable resistance, the U.S. representative at Bali agreed to significant concessions at the last minute. This made it possible to set out the Bali Roadmap for future climate control. Now the question is, will these efforts work?

To address the prospects for controlling greenhouse emissions, especially carbon dioxide, let's start with some data that reflect the views of the big players on global warming. These are the governments and interest groups with the most at stake. In all likelihood, any agreement that can be reached will be settled primarily among these few stakeholders. They include the European Union, the United States—divided between the proportion of American public opinion that favors regulating carbon dioxide and other greenhouse gas emissions and those opposed—China, and India. It also includes other relatively large economies such as Russia's, Japan's, Canada's, and Australia's, plus the growing economy of Brazil. For good measure, I have also represented environmental nongovernmental organizations (labeled here as NGOs), since they had a significant presence at Bali, and pro-environment and less sympathetic multinational corporations. In each case I have estimated potential influence in negotiations over an agreement to replace the Kyoto Protocol, position (explained in a moment), salience for mandatory emission controls, and the extent to which the stakeholder is committed to finding an agreement (even if it is not the one they favor) or will stick to their guns under political pressure (holding out for the policy they believe in). This last variable, as you know, is new to the new model I have been developing and testing for a few years. This is the model that I promised earlier I would apply to this case, just as I applied it to forecasts about Pakistan and other crises in the previous chapter.

Stakeholder	Influence	Mandatory Emission Controls	Salience	Desire for Agreement
Australia	6	65	50	50
Canada	9	60	50	50
EU	87	95	90	35
Japan	15	45	60	60
Russia	6	40	50	60
USPro	65	70	70	40
USAnti	35	30	50	30
CorpFor	3	95	50	50
CorpAgainst	3	1	75	10
NGOs	1	99	99	20
China	15	5	90	30
India	9	5	90	30
Brazil	4	3	90	40

I have rated the players' positions on a scale from 0 to 100. A position of 50 is equivalent to continuing the greenhouse gas targets that came out of the Kyoto Protocol in 1997. These standards, as I previously discussed, called for rollbacks based on 1990 emission levels. Higher values on the position scale reflect tougher standards. For example, 60 is a 10 percent toughening of standards relative to the 1990 benchmark, 100 a 50 percent increase in mandatory greenhouse emissions reductions compared to 1990. Likewise, values below 50 reflect a weakening of the terms contained in the Kyoto agreement.

Ten years passed between the Kyoto negotiations and the new round of talks that began in Bali in 2007. There were intermediate discussions in 2000 and 2001, but these were not particularly dramatic. With that in mind, I have viewed the bargaining periods as fairly long, taking exchanges of ideas among the big players about how to deal with global warming as cycling around about once every five years. That means I have simulated the negotiated standards out for about 125 years. That is certainly a long time. We will want to take more seriously the predictions closer in than farther out, since a great deal can happen between now and 2130 (when none of us will be around to check on accuracy or praise success). Because so much can happen, I have simulated the data with random shocks to salience and to each stakeholder's interest in building consensus or sticking to its guns. By randomly changing 30 percent of the salience values and 30 percent of the flexibility values in each bargaining round, we can look at a range of predicted futures to see whether the global warming simulations reveal strong trends. That will help us sort out how confident we can be about the toughness or weakness of future regulations of greenhouse gas emissions.

First let's see what the big picture looks like. Then we will examine the simulations in more detail to get a sense of how optimistic or pessimistic we should be.

The heavy solid black line in figure 11.2 shows the most likely emission standard predicted by the game. The two heavy dotted lines depict the range of regulatory values that we can be 95 percent confident includes the true future regulatory environment according to the simulations. That range of values is pretty narrow, encompassing barely five points up or down through about 2050. After that, as we should expect, there is more

FIG. 11.2. The Withering Will to Regulate Greenhouse Gases

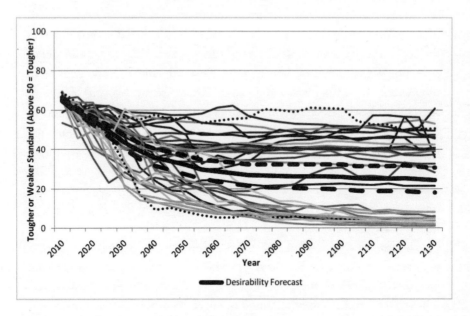

uncertainty, but even as far into the future as 2130 the range is only about ten points up or down, so these are probably pretty reliable forecasts.

The most likely value—the heavy solid line—reflects our best estimate of what the big players might broadly agree to if the global warming debate continues without any significant new discoveries in its favor or against it. It tells us two stories. First, the rhetoric of the next twenty or thirty years endorses tougher standards than those proposed—and mostly ignored—at Kyoto in 1997. We know this because the predicted value through 2025 is above 50 on the scale. That's the green part of the story. Second, support for tougher regulations falls almost relentlessly as the world closes in on 2050—a crucial date in the global warming debate. When we get to 2050, the mandatory standard being acted on is well below that set at Kyoto. By about 2070 it is down to 30, representing a significant weakening in standards. By 2100 it is closing in on 20 to 25. There's no regulatory green light left in the story by its end.

Now let's probe the details a bit. The figure shows us that there are some considerably more optimistic scenarios and also some considerably more pessimistic views that fall outside the 95 percent confidence interval. The most optimistic and pessimistic scenarios are depicted by the dotted lines at the top and bottom of the figure. The most optimistic scenario predicts no rollback in emission controls. It never dips below 50 on the scale. In fact, most of the time in this scenario the predicted level of greenhouse gas reduction hovers around 60, implying a 10 percent or so tougher standard than was agreed to in Kyoto. The pro-control faction in the United States is the driving force behind this optimistic perspective. Their salience rises from its initial level of 70 and remains remarkably high, hovering around 100. Because the issue becomes so salient to them, this U.S. group's power (resources multiplied by salience) comes to dominate debate. Although their inclination to be tough might not be enough to satisfy diehard greens, keeping this group (mostly liberal Democrats) highly engaged is the best hope for tougher standards.

Only about 10 percent of the scenarios, however, look optimistic enough to anticipate even holding the line at the standard set in the Kyoto protocol. In contrast, there are dozens of scenarios in which the standard falls close to 0, indicating abandonment of the effort to regulate greenhouse gases. Typically in these scenarios, some mix of Brazil's, India's, and China's salience rises while the salience of the pro-control faction in the United States and in the European Union drops well below their opening stance. They just seem to lose interest in greenhouse gas regulations. That decline raises its ugly head especially during global economic slowdowns, so global economic patterns are critical for us to watch as they can guide our choice of the scenarios that we should pay the most attention to. Without commitment to change by the European Union and the United States, it becomes much easier for the key developing economies to prevail with the support and even encouragement of the anticontrol American faction (mostly conservative Republicans).

Since many of my twenty-, thirty-, and even forty-year-old readers will be around in 2050, I hope you will remember to take your dusty copy of this book off the shelf then and compare the greenhouse gas predictions to the reality with which you are then living. Perhaps you'll even think to write to my children, or their children, just to say whether I got it right or wrong.

So far, there is little basis for believing greenhouse gases will be regulated away. Just in case you're still a believer in a Kyoto-style regulatory regime, but one with teeth, figure 11.3 zooms in on the biggest of the big players, at least the biggest for now: the European Union, the two U.S. factions, China, and India. Americans who worry about global warming, like their European Union brethren, remain committed through about 2030 or 2040 to tougher standards than were announced in Kyoto. But after that, they join forces with those who put economic growth ahead of regulating carbon dioxide and other emissions. We'll see shortly why that may not be so bad. The voice that dominates debate after 2040 or so is the voice of Americans who today are not convinced global warming is for real. The Chinese and the Indians support that American perspective, in the process convincing the other big players to adopt even weaker standards than those that were not enforced after Kyoto. Of course, there is little reason to think that these standards will be enforced either. I took a look at an enforcement issue, and believe me, it is not a pretty picture. No one among the real decision makers remains in favor of putting real teeth behind global climate change standards.

All of this may be leaving you rather depressed, but perhaps it shouldn't. The likely solution to global warming lies in the competitive technology

FIG. 11.3. What Will the Biggest Polluters Do About Greenhouse Gas Emissions?

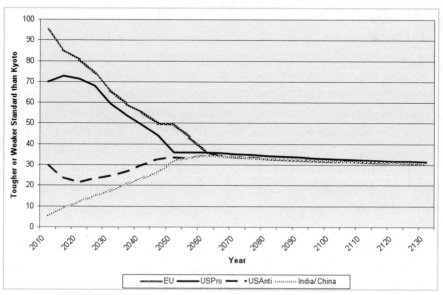

game that global warming itself helps along; it doesn't depend on the regulatory schemes that are so popular among the world's nations. These schemes, well-intentioned though they are, are also predictably vacuous. They are exercises in what game theorists call cheap talk. Promises are easily made but not easily enforced. Just look at the record of the signatories to the Kyoto Protocol.

Although the Kyoto Protocol was agreed to in December 1997, it did not take effect until February 2005. That is rather a long time for moving from agreement to presumed action on a matter of long-term global survival. Of the 175 countries, including 35 developed economies, that ratified the agreement, 137 don't have to do anything except monitor and report on their greenhouse emissions. Counted among those 137 are China, India, and Brazil. With their growing economies and their large populations, these countries are among the world's great greenhouse gas emitters. They won the battle in the negotiations that led to the Kyoto Protocol. They preserved their right to continue to pollute with no punishment for failing to do otherwise. That's what cheap talk is all about. How about Japan, one of the world's big economies that signed on to Kyoto? Remember, Japan's target is a 6 percent reduction from its 1990 emissions. The Japanese government has stated that it cannot meet its emission reduction target. Britain, while making progress on some dimensions, seems incapable of meeting its pledged reduction in carbon dioxide emissions from its 1990 level by 2010. The picture isn't pretty.

To be sure, several European Union states seem to be on track, and Russia does too, but then outside the oil sector the Russian economy has not done that well, and the Russians are only required not to increase emissions. One sure way to reduce carbon dioxide emissions is to have the economy slow down. Of course, that raises difficult political problems because people tend to vote against parties that produce poor economic performance. That could be a problem in the European Union. It's not likely to be an issue in Russia, where democracy seems to be a victim of increased oil prices. (A global economic crash, however, will bring the price of oil down, and that could jeopardize Russia's march back to autocracy.)

Anyway, what all of this amounts to is a record of cheap promises. It is easy to get governments to sign on to deals that have no teeth, no clear way to keep track of violators and to punish them. Kyoto relies heavily on self-reporting, self-policing, and goodwill. That's no way to make a global arrangement that gets its signatories to make the sacrifices needed to reduce greenhouse gas emissions.

If I sound downbeat, I am sorry. Actually, I am most optimistic for the future. My optimism, however, is despite—yup, *despite*—agreements like the ones struck in Bali or Kyoto or Copenhagen. These will be forgotten in the twinkling of an eye. They will hardly make a dent in global warming; they could even hurt by delaying serious changes. Roadmaps like the one set out at Bali make us feel good about ourselves because we did something. We looked out for future generations, we promised to do good—or did we? Unlike the pope and Holy Roman Emperor who signed on to Worms, universal schemes do not put big change into motion. Their all-inclusiveness ensures that they reflect the concerns of the lowest, not the highest, common denominator.

Deals like Bali and Kyoto include just about every country in the world. Such agreements suffer from the same wrong incentives and weak commitments as Arthur Andersen's management did in auditing Enron. To get everyone to agree to something potentially costly, the something they actually agree to must be neither very demanding nor very costly. If it is, many will refuse to join because for them the costs are greater than the benefits, or else they will join while free-riding on the costs paid by a few who were willing to bear them. That is akin to the tragedy of the commons. We all promise to protect what we hold in common—such as the earth—and then some of us cheat on the sly to enrich ourselves, figuring our little bit of cheating doesn't do any real harm. (Remember, defecting is the dominant strategy in the prisoner's dilemma.)

To get people to sign a universal agreement and not cheat, the deal must not ask them to change their behavior much from whatever they are already doing, whether that is cleaning up their neighborhood or making it dirtier. It is a race to the bottom, to the lowest common denominator. More demanding agreements weed out prospective members or encourage lies. Kyoto's demands weeded out the United States, ensuring that it could not succeed. Maybe that is what those who signed on—or at least some of them—were hoping for. They can look good and then not deliver, because after all it wouldn't be fair for them to cut back when the biggest polluter, the USA, does not.

When an agreement is demanding, lots of signatories cheat; when it isn't demanding, there is lots of compliance with what little is asked for, but then there is also little if any beneficial effect. Sacrificing self-interest for the greater good just doesn't happen very often. Governments don't throw themselves on hand grenades.[4]

It really isn't easy being green, just as Kermit the Frog has been telling us for years. Who will monitor green cheaters? The answer: interest groups, not governments; and interest groups are rarely a match for governments. Who will punish the cheaters? The answer: practically no one. The cheaters-to-be were among the rule makers when they agreed to the universal protocol. Cheating is an equilibrium strategy for many polluters, a strategy backed by the good faith and credit of their governments. Why will governments back cheaters? The answer: incentives, incentives, incentives!

Who has what incentives? There is a natural division between the rich countries whose prosperity does not depend so much on toasting our planet and the poor countries who really have no affordable alternative (yet) to fossil fuels and carbon emissions. They have an incentive to do whatever it takes to improve the quality of life of the people they govern.

The rich have an incentive to encourage the fast-growing poor to be greener, but the fast-growing poor have little incentive to listen as long as they are still poor. As the Indian government is fond of noting, sure, they are growing rapidly in income and in carbon dioxide emissions, but they are still a pale shadow of what rich countries like the United States have emitted over the centuries when they were going from poor to rich.

If the poor listen to the rich they could be in big political trouble. And when the fast-growing poor surpass the rich, the tables will turn. China, India, Brazil, and Mexico will then cry out for environmental change because that will protect their future advantaged position, while the relatively poor of that day, one or two or three hundred years from now, will resist policies that hinder their efforts to climb to the top. The rich will even fight wars to keep the rising poor from getting so rich that they threaten the old political order. (The rising poor will win those wars, by the way.)

There is also a natural division between politicians whose constituents care about the planet more than they care about their short-term quality of life—those are few and far between—and politicians whose constituents say they care about the planet but in reality often vote growth, not green. If you doubt it, take a look at the election record of green parties around the democratic world. Moreover, who will endure the political and economic costs when poor countries trot out starving children—children who would not be starving if their families could just keep on burning cow dung! We are quicker to be softhearted than we are to be green, and really, is that so bad?

So how might we solve global warming and make the world in five hundred years look attractive to our future selves? We twenty-first-century folk know of well over a hundred chemical elements and a long list of forces of nature. In Christopher Columbus's time, people pretty much only knew rain, wind, fire, and earth. They also knew hardly anything about exploiting rain, wind, and fire, but we sure do, and surely we will know more in the future. Rain, wind, and fire—they can and will solve global warming for future generations. I interpret the figures above to suggest that the reason mandatory emission standards will not be so high in 2050 is because few will care to fight that fight. It won't matter. New wind, rain, and solar technologies will be solving the problem for us.

Climate change due to global warming will add to our supply of rain, wind, and fire, and if it raises the oceans, kicks up fierce storms, and bathes us in massive quantities of BTUs, then it also adds to our urge to exploit these ancient forces just as their increased power makes us worry more. As climate change would generate more of these sources of energy, it would also create a beautiful synergy which would in turn prevent global disaster. *How could this be?*

There is an equilibrium at which enough global warming—a very modest amount more than we may already have, probably enough to be here in fifty to a hundred years (as suggested by the game's analysis)—will create enough additional sunshine in cold places, enough additional rain in dry places, enough additional wind in still places, and, most important, enough additional incentives for humankind that windmills, solar panels, hydroelectricity, and as yet undiscovered technologies will be the good, cheap, evenly distributed, and clean mechanisms to replace the fossil fuels we use today. Global warming, in other words, induces a self-solving dominant strategy in which everyone elects some mix of wind, rain, and fire technologies (and maybe even some fossil fuels in moderation) precisely because the abundance of these forces, and the attention drawn to them, will make them affordable solutions to arrest further warming—long before we all roast, drown, or are blown beyond the moon, beyond the stars, and all the way to Oz.

I am optimistic for the long future. We have already warmed enough for there to be all kinds of interesting research going on into using wind and rain and solar fire. Already there are serious discussions of solar panels and cosmic ray catchers in space and more and more windmill farms will sprout up on earth. Today such pursuits take more sacrifice than most people seem willing to make. Tomorrow that might not be true, and at that point, I doubt it'll be too late.

And, looking out five hundred years, we'll probably have figured out how to beam ourselves to distant planets where we can start all over, warming our solar system, our galaxy, and beyond with abandon.

Remember, we're looking out for numero uno.

Acknowledgments

Every author happily accumulates indebtedness. This is certainly true for me. I have been doing political forecasting and engineering for nearly thirty years cloaked in academic obscurity. Three people have been instrumental in bringing attention to my research beyond the world of political science. Michael Lerner wrote a cover story on my predictioneering for *Good* magazine. Thanks to that article, Eric Lupfer, my literary agent at the William Morris Agency, suggested that I write a book on the subject. Without Michael's article and without Eric's encouragement it would never have occurred to me that people might find this work of interest. Jonathan Jao, my editor at Random House, then turned Eric's idea and my pages into this book. I am deeply grateful to each of them for inspiring me to do this. Eric deserves additional thanks for being much more than a literary agent. He worked tirelessly at editing, rearranging, and prodding to make this a better book. I am happy to count him as my friend as well as my business associate.

My family, friends, colleagues, and students have also contributed mightily to this effort. My wife, Arlene, has given me not only the benefit of her critique of sections of this book, but her love and support for all that I do during our more than forty years together. My daughter Erin and her husband, Jason, two French horn players, not only make great music together but bring great harmony to my life. My son, Ethan, a much better applied game theorist and professor than I can ever hope to be, and his wife, Rebecca, a rabbi and an educator, add further to the harmony of my life. My daughter Gwen and her husband, Adam, two of the most fashionable and business-savvy people I know, add harmony while trying desperately—and hopelessly—to improve my sartorial splendor. They have all given me good ideas and honest feedback as this book took shape. Indeed, the title originated in a brainstorming session with Adam and Gwen, herself a terrific writer. I also owe special thanks to my sisters Mireille and Judy, two great teachers and creative spirits.

Martin Feinberg, Sam Gubins, Mary Jackman, Robert Jackman, Russell Roberts, Joseph Sherman, and Thomas Wasow discussed the ideas in this book with me and have given me the benefit of their insight and their friendship for many years. Equally, I have benefited from my friends and co-authors George Downs, James Morrow, Randolph Siverson, and Alastair Smith, who were my partners in developing some of the ideas that shape this book. Likewise, my friend and business partner, Harry Roundell, has been front and center in the analysis of many of the cases reported here and has been a source of deep and enduring support. My spring 2008 students—James Henry Ahrens, Jessica Carrano, Thomas DiLillo, Emily Leveille, Christopher Lotz, Kathryn McNish, Christian Moree, Deborah Oh, Katherine Elaine Otto, Silpa Ramineni, David Roberts, Andrea Schiferl, Jae-Hyong Shim, Jennifer Ann Thompson, Michael Vanunu, Stefan Villani, Paloma White, Natalie Wilson, Stefanie Woodburn, and Angela Zhu—and my spring 2009 students—Daniel Barker, Alexandra Bear, Katherine Cheng, Nour El-Dajani, Natalie Engdahl, Sanishya Fernando, Emily Font, Michal Harari, Andrew Hearst, Ashley Helsing, Tipper Llaguno, Veronica Mazariegos, Eric Min, Linda Moon, Shaina Negron, David Schemitsch, Kelly Siegel, Milan Sundaresan, Kenneth Villa, and Yang-Yang Zhou—served as willing guinea pigs who helped shape the chapter "Dare to Be Embarrassed!" All are innocent of responsibility for the remaining deficiencies in this book and each certainly helped eliminate many.

I am particularly grateful for the support of the Alexander Hamilton Center for Political Economy at NYU—*Predictioneer* is the embodiment of its commitment to logic and evidence in pursuit of solutions to policy problems—and to Shinasi Rama, its deputy director, and to those whose generous support makes the Center possible, especially the Veritas Fund, the Thomas W. Smith Foundation, the Lehrman Institute, David

Desrosiers, James Piereson, and especially Roger Hertog, without whom there could not be an Alexander Hamilton Center.

My colleagues in the Wilf Family Department of Politics at New York University and at the Hoover Institution at Stanford University are an endless source of support and inspiration. I could not have asked for better environments in which to pursue my research. Random House has been a terrific organization to work with, providing superb and subtle copyediting and bringing the highest standards to every aspect of this work. I thank them for their help.

Finally, I want to remember Kenneth Organski, my professor, coauthor, and friend, and the inspiration behind the original decision to apply my forecasting model to problems in the real world. He died too soon, but he left a legacy that will endure forever.

Appendix I

▪

CALCULATION OF THE WEIGHTED
MEAN PREDICTION FOR NORTH KOREA

The table below shows detailed data for some of the fifty-six stakeholders in the North Korean nuclear game, and it provides the summary values for influence times salience (that is, power) and also for influence times salience times position for all of the players. The column I × S × P is summed and divided by the sum of the column for I × S. That is, the weighted mean position equals 1,757,649 ÷ 29,384 = 59.8. This number is approximately equal to the position designated as "Slow reduction, U.S. grants diplomatic recognition."

A SAMPLE OF DATA, WITH THE CALCULATION OF THE WEIGHTED MEAN POSITION

Stakeholder	Influence	Salience	Position	I * S * P	I * S
Cho Myong Nok	4.61	90	0	0	415
Kim Chol Man	3.07	90	0	0	277
Kim Il Chol	4.61	90	0	0	415
Kim Yong Chun	4.61	90	0	0	415
North Korean Field Commanders	1.54	75	0	0	115
Paek Hak Nim	3.07	90	0	0	277
Yi Ul Sol	3.07	90	0	0	277
Yi Yong Mu	3.07	90	0	0	277
Chang Song U	3.69	90	0	0	332

Chon Pyong Ho	4.61	90	10	4150	415
J. Choso Ren	1.01	80	10	806	81
Stakeholders 12–51	...	...	...	812400	16528
Pacific Command	12.00	90	95	102600	1080
Dept. of State	22.00	95	95	198550	2090
Dept. of Defense	26.00	95	100	247000	2470
President Bush	40.00	90	100	360000	3600
United Nations	3.57	90	100	32143	321
			ISP/IS =	1757649	29384
			59.8		

Appendix II

▪

DATA USED TO ENGINEER A
COMPLEX LITIGATION

Stakeholder Bloc	Stakeholder	Influence	Salience	Position
Community	Affected Individuals	15.71	80.00	90.00
Community	Community Government	11.22	25.00	25.00
Community	Local Media	8.98	60.00	75.00
Community	National Media	6.73	25.00	25.00
Community	Plaintiff's Attorneys	5.61	80.00	100.00
Community	Union	5.61	80.00	85.00
Community	Local Experts	1.12	10.00	25.00
Congress	Ranking Democrat	8.82	40.00	75.00
Congress	Senior Democrat	7.94	30.00	25.00
Congress	Local Democrat	5.29	30.00	25.00
Congress	Ranking Republican	3.53	30.00	25.00
Congress	Local Democrat	2.65	30.00	60.00
Congress	Senior Republican	1.76	30.00	60.00
Defendant	Board of Directors	8.06	50.00	25.00
Defendant	Senior Executive	7.26	80.00	25.00
Defendant	Senior Unit Executive	7.26	80.00	25.00
Defendant	Executive	7.26	75.00	25.00
Defendant	General Counsel	3.63	75.00	25.00
Defendant	Unit President	3.63	60.00	25.00
Defendant	Unit General Counsel	2.82	80.00	25.00
Defendant	Senior Outside Attorney	1.61	80.00	25.00

Stakeholder Bloc	Stakeholder	Influence	Salience	Position
Defendant	Senior Attorney	1.21	55.00	25.00
Defendant	Unit CEO	0.40	60.00	25.00
Defendant	Group Vice President	0.40	50.00	25.00
Defendant	Corporate Committee	0.40	25.00	25.00
Defendant	Corporate Ombudsman	0.40	25.00	25.00
Defendant	Outside Attorneys	0.32	80.00	40.00
Defendant	Lead Outside Attorneys	0.32	75.00	25.00
Dept. of Justice	Deputy Attorney General	20.29	20.00	60.00
Dept. of Justice	Assistant Attorney General	16.23	50.00	75.00
Dept. of Justice	Section Chief	12.17	75.00	85.00
Dept. of Justice	Attorney	11.16	85.00	100.00
Dept. of Justice	Attorney	9.13	75.00	100.00
Dept. of Justice	Attorney	1.01	60.00	75.00
Federal Gov't	OSHA	13.64	65.00	25.00
Federal Gov't	Agency	13.64	25.00	25.00
Federal Gov't	ABC Region	2.73	20.00	80.00
State Gov't	Deputy Assistant Attorney General	4.55	20.00	75.00
State Gov't	Section Chief	3.18	20.00	75.00
State Gov't	Staff	2.27	20.00	75.00
U.S. Attorney's Ofc	U.S. Attorney	27.03	35.00	50.00
U.S. Attorney's Ofc	Office with Expertise	21.62	35.00	80.00
U.S. Attorney's Ofc	First Assistant	21.62	35.00	50.00
U.S. Attorney's Ofc	Assistant U.S. Attorney	12.16	60.00	60.00
U.S. Attorney's Ofc	FBI	8.11	20.00	60.00
U.S. Attorney's Ofc	FBI Agents	5.41	40.00	60.00
U.S. Attorney's Ofc	ABC Agents	4.05	50.00	80.00
	Initial Forecast			60.00

Notes

Introduction

1. See Bruce Bueno de Mesquita, "Leopold II and the Selectorate: An Account in Contrast to a Racial Explanation," *Historical Social Research* [*Historische Sozialforschung*] 32, no. 4 (2007): 203–21.

2. Vernon Mallinson, "Some Sources for the History of Education in Belgium," *British Journal of Educational Studies* 4, no. 1 (November 1955): 62–70.

3. See, for instance, Joseph Conrad, *Youth, and Two Other Stories* (New York: McClure, Phillips, 1903); Barbara Emerson. *Leopold II of the Belgians: King of Colonialism* (London: Weidenfield and Nicolson, 1979); Peter Forbath, *The River Congo* (New York: Harper and Row, 1977); and Adam Hochschild, *King Leopold's Ghost* (Boston: Mariner Books, 1999).

4. The discussion that follows is based on the logic and evidence provided in Bruce Bueno de Mesquita, Alastair Smith, Randolph M. Siverson, and James D. Morrow, *The Logic of Political Survival* (Cambridge, Mass.: MIT Press, 2003). See especially chapter 7. See also Bruce Bueno de Mesquita and Alastair Smith, "Political Survival and Endogenous Institutional Change," *Comparative Political Studies* 42, no. 2 (February 2009): 167–97.

5. Petty dictators typically also have a pot of money that can be spent at their sole discretion. Democratic leaders have far less authority over spending. Discretionary funds can be used to benefit the citizenry or can be socked away in a secret bank account. One way to recognize civic-mindedness is to see how many benefits the public has compared to expectations, given the type of regime. Singapore's Lee Kwan Yew and China's Deng Xiaoping, for instance, seem to have been genuinely civic-minded. They implemented effective public policies while sustaining the loyalty of their essential supporters. Kim Jong Il, Robert Mugabe, and Supreme Leader Ali Khamenei, in contrast and to varying degrees, seem not so civic-minded. See Bueno de Mesquita, Smith, Siverson, and Morrow, *Logic of Political Survival*.

6. Stanley Feder, "Factions and Policon: New Ways to Analyze Politics," in H. Bradford Westerfield, ed., *Inside CIA's Private World: Declassified Articles from the Agency's Internal Journal, 1955–1992* (New Haven: Yale University Press, 1995), and James L. Ray and Bruce M. Russett, "The Future as Arbiter of Theoretical Controversies: Predictions, Explanations and the End of the Cold War," *British Journal of Political Science* 26, no. 4 (October 1996): 441–70.

Chapter 1: What Will It Take to Put You in This Car Today?

1. See the *Jobs Rated Almanac* ratings at http://www.egguevara.com/shopping/articles/jobsrated.html.

2. If you think body language is not important, do a search online for "negotiation and body language." You will find article after article about how close sellers should place themselves to buyers, how they should use their hands and arms, facial expressions, etc. to improve the price they get and the odds of closing deals.

Chapter 2: Game Theory 101

1. John von Neumann and Oskar Morgenstern, *Theory of Games and Economic Behavior* (Princeton: Princeton University Press, 1947).

2. Sylvia Nasar, *A Beautiful Mind: The Life of Mathematical Genius and Nobel Laureate John Nash* (New York: Simon & Schuster, 1998). For progressively more thorough and technical introductions to game theory, starting with a completely readable nontechnical treatment, see Avinash K. Dixit and Barry J. Nalebuff, *Thinking Strategically: The Competitive Edge in Business, Politics and Everyday Life* (New York: W. W. Norton, 1991); James D. Morrow, *Game Theory for Political Scientists* (Princeton: Princeton University Press, 1994); Martin J. Osborne, *An Introduction to Game Theory* (Oxford: Oxford University Press, 2003); and Drew Fudenberg and Jean Tirole, *Game Theory* (Cambridge, Mass.: MIT Press, 1991).

3. Six dollars per day falls in the middle of the World Bank's estimate of per capita income for Iraq in 2007. Unlike for most countries, for Iraq the World Bank is not able to provide a firm number. Other estimates seem to be based on the World Bank's range.

4. Brian Kolodiejchuk, ed., *Mother Teresa: Come Be My Light—The Private Writings of the Saint of Calcutta* (New York: Doubleday, 2007).

5. See Irene Hau-siu Chow, Victor P. Lau, Thamis Wing-chun Lo, Zhenquan Sha, and He Yun, "Service Quality in Restaurant Operations in China: Decision- and Experiential-Oriented Perspectives," *International Journal of Hospitality Management* 26, no. 3 (September 2007): 698–710.

6. For the math mavens out there, the circle is a special case in which each dimension is of equal importance to the player. If one dimension is more important than the other, then we should draw an ellipse, each of whose radii reflects the relative importance of the issues. I skip this complicating detail here.

7. Those mathematically inclined and interested in delving more deeply into how it is possible for rational decision makers to move from any policy combination to any other, see Richard McKelvey, "Intransitivities in Multidimensional Voting Models and Some

Implications for Agenda Control," *Journal of Economic Theory* 12 (1976): 472–82; Richard McKelvey, "General Conditions for Global Intransitivities in Formal Voting Models," *Econometrica* 47 (1979): 1085–1112; and Norman Schofield, "Instability of Simple Dynamic Games," *Review of Economic Studies* 45 (1978): 575–94.

8. For a careful study of steroid use that is broadly consistent with the values used in this example, see Jenny Jakobsson Schulze, Jonas Lundmark, Mats Garle, Ilona Skilving, Lena Ekström, and Anders Rane, "Doping Test Results Dependent on Genotype of Uridine Diphospho-Glucuronosyl Transferase 2B17, the Major Enzyme for Testosterone Glucuronidation," *Journal of Clinical Endocrinology & Metabolism* 93, no. 7 (July 2008): 2500–2506. Based on the values in this study, about 9 percent of a random sample from the population would falsely test positive. I assume 10 percent.

9. Bayes' Theorem allows us to answer the question "What is the probability a person is of a particular type (such as a performance enhancing steroid user) given that they say or do something in particular (such as test positive for steroids)?" To answer this question we must solve the following calculation: Let P stand for probability, R for being a steroid user, S for testing positive, and ~R for being the type of baseball player who does not use steroids. The straight line symbol | is read as "given." Then $P(R \mid S) = \frac{P(S \mid R)P(R)}{P(S \mid R)P(R) + P(S \mid -R)P(-R)}$ is read as "the probability of being a steroid user given that you tested positive ($P[R \mid S]$) equals the probability of testing positive given that you are a steroid user times the probability of being a steroid user divided by that same quantity plus the probability of testing positive given that you are not a steroid user times the probability that you are not a steroid user." Thus, the calculation is conditioned on the two sets of people who test positive: those who use steroids and those who don't. In the baseball example this translates into $P(R \mid S) = \frac{(.9)(.1)}{(.9)(.1) + (.1)(.9)} = \frac{.09}{.18} = .5$.

Chapter 3: Game Theory 102

1. The first major efforts to show that arms races lead to war are the work of Lewis Fry Richardson, a distinguished meteorologist who predicted World War I but, using the same logic, failed to anticipate World War II. See his *Arms and Insecurity* (Chicago: Quadrangle, 1960). The literature tying arms races to war is vast but almost universally fails to consider that arms purchases are anticipatory, or, in game-theory lingo, they are endogenous, strategic decisions.

2. The subject of renegotiation-proofness has attracted the interest of many economists, leading to a vast literature. Some seminal papers include Dilip Abreu, David Pearce, and Ennio Stacchetti, "Renegotiation and Symmetry in Repeated Games," *Journal of Economic Theory* 60, no. 2 (1993): 217–40; Jean-Pierre Benoit and Vijay Krishna, "Renegotiation in Finitely Repeated Games," *Econometrica* 61 (1993): 303–23; and James Bergin and W. Bentley MacLeod, "Efficiency and Renegotiation in Repeated Games," *Journal of Economic Theory* 61, no. 1 (1993): 42–73.

3. The seminal work on this question dates back to the eighteenth-century French philosopher, mathematician, and nobleman the Marquis de Condorcet. Regrettably, the latter characteristic—he opposed beheading the king and queen—cost him his life during the French Revolution. There is a wonderful statue of him on the left bank of the Seine, not too far from Notre Dame. I always pay homage to him when I am in Paris. His insights were built upon in the second half of the twentieth century to establish the

modern understanding of voting methods. See, for instance, Kenneth Arrow, *Social Choice and Individual Values* (New York: John Wiley and Sons, 1951); William H. Riker, *Liberalism Against Populism* (New York: Freeman, 1982); Richard D. McKelvey and Norman Schofield, "Structural Instability of the Core," *Journal of Mathematical Economics* 15, no. 3 (June 1986): 179–98; and Gary W. Cox, *Making Votes Count* (New York: Cambridge University Press, 1997).

Chapter 4: Bombs Away

1. Stanley Feder, "Factions and Policon: New Ways to Analyze Politics," in H. Bradford Westerfield, ed., *Inside CIA's Private World: Declassified Articles from the Agency's Internal Journal, 1955–1992* (New Haven: Yale University Press, 1995).

2. This is a casual statement of the median voter theorem, one of the most important concepts in understanding issue resolutions. See Duncan Black, *The Theory of Committees and Elections* (Cambridge: Cambridge University Press, 1958), and Anthony Downs, *An Economic Theory of Democracy* (New York: Harper, 1957).

3. This second first-cut prediction relies on the mean voter theorem. See Andrew Caplin and Barry Nalebuff, "Aggregation and Social Choice: A Mean Voter Theorem," *Econometrica* 59 (1991): 1–23; and Norman Schofield, "The Mean Voter Theorem: Necessary and Sufficient Conditions for Convergent Equilibrium," *Review of Economic Studies* 74 (2007): 965–80.

4. See Bruce Bueno de Mesquita, "Ruminations on Challenges to Prediction with Rational Choice Models," *Rationality and Society* 15, no. 1 (2003): 136–47; and Robert Thomson, Frans N. Stokman, and Christopher H. Achen, eds., *The European Union Decides* (Cambridge: Cambridge University Press, 2006).

Chapter 5: Napkins for Peace

1. For a not-too-technical explanation of what goes on inside the model, see Bruce Bueno de Mesquita, *Predicting Politics* (Columbus: Ohio State University Press, 2002).

2. Bruce Bueno de Mesquita, "Multilateral Negotiations: A Spatial Analysis of the Arab-Israeli Dispute," *International Organization* (Summer 1990): 317–40.

3. See http://news.bbc.co.uk/1/hi/world/middle_east/1763912.stm.

4. Anthony H. Cordesman, *The Israel-Palestinian War: Escalating to Nowhere* (Westport, Conn.: Praeger, 2005): 219.

Chapter 6: Engineering the Future

1. For an early, foundational study applying war-of-attrition games, see John Maynard Smith and Geoffrey A. Parker, "The Logic of Asymmetric Contests," *Animal Behaviour* 24 (1976): 159–75. See also Anatol Rapaport, *Two Person Game Theory* (Ann Arbor: University of Michigan Press, 1966).

2. See, for instance, James D. Fearon, "Rationalist Explanations for War," *International Organization* 49 (1995): 379–414.

Chapter 7: Fast-Forward the Present

1. For a deeper exploration of some surprising implications of commitment problems and the pursuit of peace with adversaries, in particular peace with terrorists, see Ethan Bueno de Mesquita, "Conciliation, Counter-Terrorism, and Patterns of Terrorist Violence," *International Organization* 59, no. 1 (2005): 145–76.

2. The actual calculation predicting the impact of deaths (the horizontal axis) on tourism (the vertical axis) is based on the logarithm of deaths to capture order-of-magnitude changes. Doubling the lives lost from 10 deaths to 20, for instance, represents a more noticeable change than going from 190 to 200 deaths even though the absolute change is the same. Logarithms capture the magnitude of change so that equal spaces reflect equal percentage increments in lost lives.

3. Limitations on the availability of data constrain the number of years I can cover. Violence and tourism are both measured quarterly. The Palestinians suffer the lion's share of violent deaths. The graph looks very much the same if only Palestinian deaths are plotted rather than all violent deaths resulting from the conflict. Data on Israeli tourism are from the Bank of Israel and can be found at http://www.bankisrael.gov.il/series/en/catalog/tourism/tourist entries/. Data on violent deaths are from David Fielding, "How Does Violent Conflict Affect Investment Location Decisions?" *Journal of Peace Research* 41, no. 4 (2004): 465–84.

4. See "Palestinian Central Bureau of Statistics Press Release for the Hotel Survey, Fourth Quarter 2005," found at www.pcbs.pna.org/Portals/_pcbs/PressRelease/HOTEL0405.pdf.

Chapter 8: How to Predict the Unpredictable

1. See John Lewis Gaddis, "International Relations Theory and the End of the Cold War," *International Security* 17, no. 3 (Winter 1992): 323–73; and James Ray and Bruce Russett, "The Future as Arbiter of Theoretical Controversies: Predictions, Explanations and the End of the Cold War," *British Journal of Political Science* 26, no. 4 (October 1996): 441–70.

2. Bruce Bueno de Mesquita, "Measuring Systemic Polarity," *Journal of Conflict Resolution* (June 1975): 187–215; and Michael F. Altfeld and Bruce Bueno de Mesquita, "Choosing Sides in Wars," *International Studies Quarterly* (March 1979): 87–112.

3. EUGene's data can be accessed at http://www.eugenesoftware.org/.

4. See Bruce Bueno de Mesquita, "The End of the Cold War: Predicting an Emergent Property," *Journal of Conflict Resolution* 42, no. 2 (April 1998): 131–55.

Chapter 9: Fun with the Past

1. See Xenophon, *Hellenica,* Book VI, Chapter IV, downloaded from http://www.fordham.edu/halsall/ancient/371leuctra.html).

2. Edward Kritzler, *Jewish Pirates of the Caribbean: How a Generation of Swashbuckling Jews Carved Out an Empire in the New World in Their Quest for Treasure, Religious Freedom—and Revenge* (New York: Doubleday, 2008).

3. Niall Ferguson, *The Pity of War: Explaining World War I* (New York: Basic Books, 2000).

Chapter 10: Dare to Be Embarrassed!

1. See my TED talk at http://ow.ly/2gFz for my predictions about Iran's nuclear program.

2. The Algiers Accord, signed in 1975, was supposed to have resolved Iraq-Iran border disputes, such as over control of the Shatt al-Arab River near Basra where the river defines the boundary between Iran and Iraq. Despite the agreement, Saddam Hussein's Iraq attacked Iran six years later, beginning a war that lasted for eight years. The border area remains a source of conflict while the Algiers Accord provides the legally—but not strategically—binding terms for delineating the boundaries of Iran and Iraq. As we know, promises are not the same as commitments, and nothing could be truer in the history of relations between Iran and Iraq.

3. A related analysis indicates that Tehran's mayor, Ghalibaf. The next Iranian presidential election, though looking very close in my analyses, seems to give a small edge to Ahmadinejad. His power decline is more substantial after the election than before.

4. The patterns of political change shown here arise as well in analysis I did using completely different data on Iran. That gives me considerable confidence that they accurately reflect changing influence in Iran. For a portion of those other analyses see my TED talk at http://ow.ly/2gFz.

Chapter 11: The Big Sweep

1. The text of Lateran II can be found at www.fordham.edu/halsall/basis/lateran2.html.

2. Raoule Van Caenegem, *The Birth of the English Common Law* (Cambridge: Cambridge University Press, 1988): 64.

3. Emmanuel Le Roy Ladurie, *Times of Feast, Times of Famine* (New York: Doubleday, 1971).

4. George W. Downs, David M. Rocke, and Peter Barsoom, "Is the Good News About Compliance Good News for Cooperation," *International Organization* 50 (1996): 379–406.

Index

Page numbers in *italics* refer to figures and tables.